A SHEPHERD'S LIFE

A SHEPHERD'S LIFE

W. H. HUDSON

WITH WOOD ENGRAVINGS BY
REYNOLDS STONE

COMPTON PRESS

First edition 1910
This edition © The Compton Press 1978

Set in Linotype Pilgrim
designed by Humphrey Stone
printed and published 1978 by The Compton Press Ltd.
The Old Brewery, Tisbury, Wiltshire

ISBN 0 900193 50 6

Reprinted 1979

CONTENTS

FOREWORD

IT WAS the fashion in the 1920s for literary men to speak of Joseph Conrad and William Henry Hudson as the twin peaks of English writing, the successors of Meredith and Hardy. It has never been fashionable to say much about them as exotic transplants into the English literary scene—which they both were—or to point out that neither of them was widely appreciated during the greater part of his working life. Interest in Conrad was revived after the Second World War when Dent issued a collected edition of his works and he found a wide and appreciative audience. A start was made by the same publisher on a cheap uniform edition of Hudson's works but only seven volumes were issued, presumably through lack of public interest. Evidently Hudson's star was not in the ascendant.

Today there is a new and vital interest in the works of Hudson, a man who never became a cult figure in his own lifetime but who could so easily—and perhaps deservedly—become one now. We are in a much better position than were his contemporaries to notice and appreciate the things he wrote about and to sympathise with the kind of practical improvements he campaigned for so vigorously. His concern for the welfare of his beloved birds, for instance, was not limited to writing about them in book after book but also found expression in a more personal involvement such as the preliminary skirmishing which resulted in the formation of the Royal Society for the Protection of Birds, an organisation which virtually recognises Hudson as its patron saint.

Initially Hudson hoped to become famous, or at least hoped to secure a modest living, by writing novels but when his first two books failed to make any meaningful impression on the reading public he turned to the manufacture of what is rather inadequately called 'nature writing'. In 1892 he produced *The Naturalist in La Plata*, based on his own experiences in the pampas, which was the first of his books to achieve a moderate success, though more from a scientific than a literary point of view, the illustrious naturalist Alfred Russell Wallace having given it a long and glowing review in *Nature*. His first commercial success, however, was *Green Mansions*, a romantic novel first published in 1904. A hauntingly tragic story set in the South American jungle it was not until 1916, the year of Alfred Knopf's revised American edition with an introductory essay by John Galsworthy, that Hudson began to make an appreciable amount of money out of his writing.

Green Mansions is still Hudson's most popular book and many of his admirers would say it is his best. But in 1910 he wrote a very different book, one which may be considered his most consistent and most sustained literary performance: *A Shepherd's Life*. In no sense a novel— though many of the names in it are fictional—it bears some resemblance to his *Nature in Downland*, published a decade earlier, each having a downland setting and each being full of interesting and well-reported information about country life. But there the resemblance ends. *Nature in Downland* is a very good book: *A Shepherd's Life* is a classic.

Emphatically a book about men rather than sheep Hudson tells us, in his delightful chapter on gypsies, how a large part of the material composing it was obtained. 'It came to me in conversations, at intervals', he says, 'during several years with the shepherd. In his long life in his native village, a good deal of it spent on the quiet down, he had seen many things it was or would be interesting to hear; the things which had interested him, too, at the time, and had fallen into oblivion, yet might be recovered'. To a contemplative, patient man such as Hudson the long process of extracting information from the 'very tall, big-boned, lean, round-shouldered man . . . uncouth almost to the verge of grotesqueness' was a congenial exercise. 'When I turned into the shepherd's cottage', he says, 'if it was in winter and he was sitting by the fire, I would sit and smoke with him, and if we were in a talking mood I would tell him where I had been and what I had heard and seen, on the heath, in the woods, in the village, or anywhere, on the chance of its reminding him of something worth hearing in his past life'.

In this leisurely way Hudson acquired the raw materials from which he fashioned the story of Caleb Bawcombe, the shepherd of Winterbourne Bishop, and the rural community in south-eastern Wiltshire to which he belonged and from which he never strayed. Hudson never disclosed the real name of the shepherd or that of the village. Later research has shown that Caleb Bawcombe, who as Hudson tells us 'was lame for life', may be identified conclusively with one James Lawes, who is known to have walked with a limp. Winterbourne is the fictional alias of Martin, a village on the borders of Wiltshire, Hampshire and Dorset. Undoubtedly both James Lawes and Martin gain rather than lose by being translated to *A Shepherd's Life* for it is doubtful if Hudson could have written at such length about a man or a place if he was not predisposed to like them and to show them in a favourable light. It is not surprising that the man who preferred a solitary sparrow to the whole of London's teeming masses should write a book about the life of a shepherd. A naturalist in the broadest sense Hudson took all nature as his province and that encompassed humble human beings as well as birds.

A Shepherd's Life is primarily about men and women. Perhaps that is why it succeeds so well as a book. For most of us animal lore cannot rival the natural history of humans in interest. As Hudson was a born story teller and his books abound in stories based on a mixture of truth and imagination it would be folly to describe the book as a faithful account of old Wiltshire life. But if he embroidered any of the tales he tells to make *A Shepherd's Life* more cohesive and readable it is difficult to discover where truth gives way to fiction, so perfect and so plausible is the blending.

Here, as in no other book he had written before, Hudson's indignation against man's frequently inhuman treatment of his fellows is expressed forcibly, particularly so in the two chapters entitled 'Old Wiltshire Days'. Angrily he tells of men forced to filch a swede or two from the fields because they had no work and were starving, of those whose hunger would not be satisfied with anything less than mutton, and of 'the "human devil" in a black cap', as Hudson called him, who would send them to the gallows or to Botany Bay for necessarily stealing the sheep which provided the mutton. It mattered not that such things happened nearly a century earlier: time had no meaning for Hudson and could not wash away the facts of human suffering and injustice.

Little is said about sheep but a great deal about sheep dogs. Probably Hudson found sheep uninteresting as objects of study but he was fascinated by the intelligence of the sheep dog and the mysterious relationship between it and its master. As might have been expected the old shepherd told him numerous stories about sheep dogs. Two entire chapters are devoted to them and particularly to one of Caleb's own dogs, a remarkable animal who could do almost anything with sheep, played with rabbits without harming them, nearly died from the effects of an adder's bite, and ended his life of unswerving devotion and service in the usual, accepted and most merciful way: by the gun.

Birds crop up here and there throughout the book but they were not important enough to Caleb for him to say a lot about them, and Hudson—with admirable restraint—does not have him say more than he ought. He writes of the ravens which used to be attracted to the half-consumed and putrefying carcasses of horses stripped of their hides and left in the woods for the benefit of the deer hounds. Large numbers of these carrion-eating birds congregated to enjoy their grisly meals and it was only with the decline of deer hunting as a sport that the ravens began to disappear from Wiltshire. The commonest bird by far, the starling, was Caleb's favourite, the one bird constantly associated with sheep in the pasture. A wise bird, it would ignore the kestrel 'but if a sparrowhawk made its appearance, instantly the crowd of birds could be seen flying at furious speed towards the nearest flock of sheep and down into the flock they would fall like a shower of stones and instantly disappear from sight'.

Towards the end of the book Hudson says that the apparently scanty harvest of facts about animal life, wild and domestic, he had gathered in his talks with the old shepherd seemed, in his opinion, 'a somewhat abundant one'. Compared to other men in his microcosm Caleb was observant and had sympathy for the animals he observed. He was immeasurably more observant and sympathetic, for instance, than the ubiquitous gamekeeper, a licensed killer of animals whose way of life brought endless opportunities for observation if not sympathy. In truth there was not a great deal more Caleb could have observed, his tasks having kept him to fields and downs where wild life was least abundant and varied.

On the other hand was there much more to be seen in a wood, the gamekeeper's traditional preserve? For answer Hudson described his experiences in the Great Ridge Wood overlooking the vale of the Wylye. On his second day in it he startled a cock pheasant which, in turn, disturbed a roe-deer. But the naturalist's expectations of the many delights which lay in store for him on future occasions were not fulfilled. 'That was the best and the only great thing I saw in the Great Ridge Wood', he says, 'for the curse of the pheasant is on it as on all the woods and forests in Wiltshire, and all wild life considered injurious to the semi-domestic bird, from the sparrowhawk to the harrier and buzzard and goshawk, and from the little mousing weasel to the badger; and all the wild life that is only beautiful, or which delights us because of its wildness, from the squirrel to the roe-deer, must be included in the slaughter'. Oh how Hudson hated the gamekeeper, the man whose job it was to shoot any wild thing which may conceivably harm the pheasant, the man who ensured that there would be a plentiful supply of that gaudy and succulent creature for other, more fastidious gunmen to blast out of the sky! And undoubtedly he is at his best as a writer when he is indignant. It is then that his words acquire an edge, a power which his purer nature writings never have. It is then that he becomes a compelling writer.

Fortunately there has been a great improvement in the relationship between men and animals in most of the places where Hudson used to keep vigil; and the rough justice which prevailed among men when Caleb was young has long since been softened. Little of the old order remains but as long as we have *A Shepherd's Life* we shall never entirely lose sight of it. Perhaps, after all, that is why Hudson wrote it and why he wrote it so well.

PETER DANCE

I

SALISBURY PLAIN

WILTSHIRE looks large on the map of England, a great green county, yet it never appears to be a favourite one to those who go on rambles in the land. At all events I am unable to bring to mind an instance of a lover of Wiltshire who was not a native or a resident, or had not been to Marlborough and loved the country on account of early associations. Nor can I regard myself as an exception, since, owing to a certain kind of adaptiveness in me, a sense of being at home wherever grass grows, I am in a way a native too. Again, listen to any half-dozen of your friends discussing the places they have visited, or intend visiting, comparing notes about the counties, towns, churches, castles, scenery—all that draws them and satisfies their nature, and the chances are that they will not even mention Wiltshire. They all know it 'in a way'; they have seen Salisbury Cathedral and Stonehenge, which everybody must go to look at once in his life; and they have also viewed the country from the windows of a railroad carriage as they passed through on their flight to Bath and to Wales with its mountains, and to the west country, which many of us love best of all—Somerset, Devon, and Cornwall. For there is nothing striking in Wiltshire, at all events to those who love nature first; nor mountains, nor sea, nor anything to compare with the places they are hastening to, west or north. The downs! Yes, the downs are there, full in sight of your window, in their flowing forms resembling vast, pale green waves, wave beyond wave, 'in fluctuation fixed'; a fine country to walk on in fine weather for all those who regard the mere exercise of walking as sufficient pleasure. But to those who wish for something more, these downs may be neglected, since, if downs are wanted, there is the higher, nobler Sussex range within an hour of London. There are others on whom the naked aspect of the downs has a repelling effect. Like Gilpin they love not an undecorated earth; and false and ridiculous as Gilpin's taste may seem to me and to all those who love the chalk, which 'spoils everything' as Gilpin said, he certainly expresses a feeling common to those who are unaccustomed to the emptiness and silence of these great spaces.

As to walking on the downs, one remembers that the fine days are not so many, even in the season when they are looked for—they have certainly been few during this wet and discomfortable one of 1909. It is indeed only on the chalk hills that I ever feel disposed to quarrel with this English climate, for all weathers are good to those who love the open air, and have their special attractions. What a pleasure it is to be out in rough weather in October when the equinoctial gales are on, 'the wind Euroclydon,' to listen to its roaring in the bending trees, to watch the dead leaves flying, the pestilence-stricken multitudes, yellow and black and red, whirled away in flight on flight before the volleying blast, and to hear and see and feel the tempests of rain, the big silver-grey drops that smite you like hail! And what pleasure too, in the still grey November weather, the time of suspense and melancholy before winter, a strange quietude, like a sense of apprehension in nature! And so on through the revolving year, in all places in all weathers, there is pleasure in the open air, except on these chalk hills because of their bleak nakedness. There the wind and driving rain are not for but against you, and may overcome you with misery. One feels their loneliness, monotony, and desolation on many days, sometimes even when it is not wet, and I here recall an amusing encounter with a bird-scarer during one of these dreary spells.

It was in March, bitterly cold, with an east wind which had been blowing many days, and overhead the sky was of a hard, steely grey. I was cycling along the valley of the Ebble, and finally leaving it pushed up a long steep slope and set off over the high plain by a dusty road with the wind hard against me. A more desolate scene than the one before me it would be hard to imagine, for the land was all ploughed and stretched away before me, an endless succession of vast grey fields, divided by wire fences. On all that space there was but one living thing in sight, a human form, a boy, far away on the left side, standing in the middle of a big field with something which looked like a gun in his hand. Immediately after I saw him he, too, appeared to have caught sight of me, for turning he set off running as fast as he

2

could over the ploughed ground towards the road, as if intending to speak to me. The distance he would have to run was about a quarter of a mile and I doubted that he would be there in time to catch me, but he ran fast and the wind was against me, and he arrived at the road just as I got to that point. There by the side of the fence he stood, panting from his race, his handsome face glowing with colour, a boy about twelve or thirteen, with a fine strong figure, remarkably well dressed for a bird-scarer. For that was what he was, and he carried a queer, heavy-looking old gun. I got off my wheel and waited for him to speak, but he was silent, and continued regarding me with the smiling countenance of one well pleased with himself. 'Well?' I said, but there was no answer; he only kept on smiling.

'What did you want?' I demanded impatiently.

'I didn't want anything.'

'But you started running here as fast as you could the moment you caught sight of me.'

'Yes, I did.'

'Well, what did you do it for—what was your object in running here?'

'Just to see you pass,' he answered.

It was a little ridiculous and vexed me at first, but by and by when I left him, after some more conversation, I felt rather pleased; for it was a new and somewhat flattering experience to have any person run a long distance over a ploughed field, burdened with a heavy gun, 'just to see me pass.'

But it was not strange in the circumstances; his hours in that grey, windy desolation must have seemed like days, and it was a break in the monotony, a little joyful excitement in getting to the road in time to see a passer-by more closely, and for a few moments gave him a sense of human companionship. I began even to feel a little sorry for him, alone there in his high, dreary world, but presently thought he was better off and better employed than most of his fellows poring over miserable books in school, and I wished we had a more rational system of education for the agricultural districts, one which would not keep the children shut up in a room during all the best hours of the day, when to be out of doors, seeing, hearing, and doing, would fit them so much better for the life-work before them. Squeers' method was a wiser one. We think less of it than of the delightful caricature, which makes Squeers 'a joy for ever,' as Mr. Lang has said of

3

Pecksniff. But Dickens was a Londoner, and incapable of looking at this or any other question from any other than the Londoner's standpoint. Can you have a better system for the children of all England than this one which will turn out the most perfect draper's assistant in Oxford Street, or, to go higher, the most efficient Mr. Guppy in a solicitor's office? It is true that we have Nature's unconscious intelligence against us; that by and by, when at the age of fourteen the boy is finally released, she will set to work to undo the wrong by discharging from his mind its accumulations of useless knowledge as soon as he begins the work of life. But what a waste of time and energy and money! One can only hope that the slow intellect of the country will wake to this question some day, that the countryman will say to the townsman, Go on making your laws and systems of education for your own children, who will live as you do indoors; while I shall devise a different one for mine, one which will give them hard muscles and teach them to raise the mutton and pork and cultivate the potatoes and cabbages on which we all feed.

To return to the downs. Their very emptiness and desolation, which frightens the stranger from them, only serves to make them more fascinating to those who are intimate with and have learned to love them. That dreary aspect brings to mind the other one, when, on waking with the early sunlight in the room, you look out on a blue sky, cloudless or with white clouds. It may be fancy, or the effect of contrast, but it has always seemed to me that just as the air is purer and fresher on these chalk heights than on the earth below, and as the water is of a more crystal purity, and the sky perhaps bluer, so do all colours and all sounds have a purity and vividness and intensity beyond that of other places. I see it in the yellows of hawkweed, rock-rose, and birds'-foot-trefoil, in the innumerable specks of brilliant colour—blue and white and rose—of milk-wort and squinancy-wort, and in the large flowers of the dwarf thistle, glowing purple in its green setting; and I hear it in every bird-sound, in the trivial songs of yellow-hammer and corn-bunting, and of dunnock and wren and whitethroat.

The pleasure of walking on the downs is not, however, a subject which concerns me now; it is one I have written about in a former work, 'Nature in Downland,' descriptive of the South Downs. The theme of the present work is the life, human and other, of the South Wiltshire Downs, or of Salisbury Plain. It is the part of Wiltshire which has most attracted me. Most persons would say that the Marlborough Downs are greater, more like the great Sussex range as it appears from the Weald: but chance brought me farther south, and the character and life of the village people when I came to know them made this appear the best place to be in.

The Plain itself is not a precisely defined area, and may be made to include as much or little as will suit the writer's purpose. If you want a continuous plain, with no dividing valley cutting through it, you must place it between the Avon and Wylye Rivers, a distance about fifteen miles broad and as many long, with the village of Tilshead in its centre; or, if you don't mind the valleys, you can say it extends from Downton and Tollard Royal south of Salisbury to the Pewsey vale in the north, and from the Hampshire border on the east side to Dorset and Somerset on the west, about twenty-five to thirty miles each way. My own range is over this larger Salisbury Plain, which includes the River Ebble, or Ebele, with its numerous interesting villages, from Odstock and Combe Bisset, near Salisbury and 'the Chalks,' to pretty Alvediston near the Dorset line, and all those in the Nadder valley,

and westward to White Sheet Hill above Mere. You can picture this high chalk country as an open hand, the left hand, with Salisbury in the hollow of the palm, placed nearest the wrist, and the five valleys which cut through it as the five spread fingers, from the Bourne (the little finger) succeeded by Avon, Wylye, and Nadder, to the Ebble, which comes in lower down as the thumb and has its junction with the main stream below Salisbury.

A very large portion of this high country is now in a transitional state, that was once a sheep-walk and is now a training ground for the army. Where the sheep are taken away the turf loses the smooth elastic character which makes it better to walk on than the most perfect lawn. The sheep fed closely, and everything that grew on the down—grasses, clovers, and numerous small creeping herbs—had acquired the habit of growing and flowering close to the ground, every species and each individual plant striving, with the unconscious intelligence that is in all growing things, to hide its leaves and pushing sprays under the others, to escape the nibbling teeth by keeping closer to the surface. There are grasses and some herbs, the plantain among them, which keep down very close but must throw up a tall stem to flower and seed. Look at the plantain when its flowering time comes; each particular plant growing with its leaves so close down on the surface as to be safe from the busy, searching mouths, then all at once throwing up tall, straight stems to flower and ripen its seeds quickly. Watch a flock at this time, and you will see a sheep walking about, rapidly plucking the flowering spikes, cutting them from the stalk with a sharp snap, taking them off at the rate of a dozen or so in twenty seconds. But the sheep cannot be all over the downs at the same time, and the time is short, myriads of plants throwing up their stems at once, so that many escape, and it has besides a deep perennial root so that the plant keeps its own life though it may be unable to sow any seeds for many seasons. So with other species which must send up a tall flower stem; and by and by, the flowering over and the seeds ripened or lost, the dead, scattered stems remain like long hairs growing out of a close fur. The turf remains unchanged; but take the sheep away and it is like the removal of a pressure, or a danger: the plant recovers liberty and confidence and casts off the old habit; it springs and presses up to get the better of its fellows—to get all the dew and rain and sunshine that it can—and the result is a rough surface.

Another effect of the military occupation is the destruction of the wild life of the Plain, but that is a matter I have written about in my last book, 'Afoot in England,' in a chapter on Stonehenge, and need not dwell on here. To the lover of Salisbury Plain as it was, the sight of military camps, with white tents or zinc houses, and of bodies of men in khaki marching and drilling, and the sound of guns, now informs him that he is in a district which has lost its attraction, where nature has been dispossessed.

Meanwhile, there is a corresponding change going on in the human life of the district. Let anyone describe it as he thinks best, as an improvement or a deterioration, it is a great change nevertheless, which in my case and probably that of many others is as disagreeable to contemplate as that which we are beginning to see in the down, which was once a sheep-walk and is so no longer. On this account I have ceased to frequent that portion of the Plain where the War Office is in possession of the land, and to keep to the southern side in my rambles, out of sight and hearing of the 'white-tented camps' and mimic warfare. Here is Salisbury Plain as it has been these thousand years past, or ever since sheep were pastured here more than in any other district in England, and that may well date even more than ten centuries back.

Undoubtedly changes have taken place even here, some very great, chiefly during the last, or from the late eighteenth century. Changes both in the land and the animal life, wild and domestic. Of the losses in wild bird life there will be something to say in another chapter; they relate chiefly to the extermination of the finest species, the big bird, especially the soaring bird, which is now gone out of all this wide Wiltshire sky. As a naturalist I must also lament the loss of the old Wiltshire breed of sheep, although so long gone. Once it was the only breed known in Wilts, and extended over the entire county; it was a big animal, the largest of the fine-wooled sheep in England, but for looks it certainly compared badly with modern downland breeds and possessed, it was said, all the points which the breeder, or improver, was against. Thus, its head was big and clumsy, with a round nose, its legs were long and thick, its belly without wool, and both sexes were horned. Horns, even in a ram, are an abomination to the modern sheep-farmer in Southern England. Finally, it was hard to fatten. On the other hand it was a sheep which had been from of old on the bare open downs and was modified to suit the conditions, the scanty feed, the bleak, bare country, and the long distances it had to travel to and from the pasture ground. It was a strong, healthy, intelligent animal, in appearance and character like the old original breed of sheep on the pampas of South America, which I knew as a boy, a coarse-wooled sheep with naked belly, tall and hardy, a greatly modified variety of the sheep introduced by the Spanish colonist three centuries ago. At all events the old Wiltshire sheep had its merits, and when the South Down breed was introduced during the late eighteenth century the farmer viewed it with disfavour; they liked their old na-

7

tive animal, and did not want to lose it. But it had to go in time, just as in later times the South Down had to go when the Hampshire Down took its place—the breed which is now universal, in South Wilts at all events.

A solitary flock of the pure-bred old Wiltshire sheep existed in the county as late as 1840, but the breed has now so entirely disappeared from the country that you find many shepherds who have never even heard of it. Not many days ago I met with a curious instance of this ignorance of the past. I was talking to a shepherd, a fine intelligent fellow, keenly interested in the subjects of sheep and sheep-dogs, on the high down above the village of Broad Chalk on the Ebble, and he told me that his dog was of mixed breed, but on its mother's side came from a Welsh sheep-dog, that his father had always had the Welsh dog, once common in Wiltshire, and he wondered why it had gone out as it was so good an animal. This led me to say something about the old sheep having gone out too, and as he had never heard of the old breed I described the animal to him.

What I told him, he said, explained something which had been a puzzle to him for some years. There was a deep hollow in the down near the spot where we were standing, and at the bottom he said there was an old well which had been used in former times to water the sheep, but masses of earth had fallen down from the sides, and in that condition it had remained for no one knew how long—perhaps fifty, perhaps a hundred years. Some years ago it came into his master's head to have this old well cleaned out, and this was done with a good deal of labour, the sides having first been boarded over to make it safe for the workmen below. At the bottom of the well a vast store of rams' horns was discovered and brought out; and it was a mystery to the farmer and the men how so large a number of sheep's horns had been got together; for rams are few and do not die often, and here there were hundreds of horns. He understood it now, for if all the sheep, ewes as well as rams, were horned in the old breed, a collection like this might easily have been made.

The greatest change of the last hundred years is no doubt that which the plough has wrought in the aspect of the downs. There is a certain pleasure to the eye in the wide fields of golden corn, especially of wheat, in July and August; but a ploughed down is a down made ugly, and it strikes one as a mistake, even from a purely economic point of view, that this old rich turf, the slow product of centuries, should be ruined for ever as sheep-pasture when so great an extent of uncultivated land exists elsewhere, especially the heavy clays of the Midlands, better suited for corn. The effect of breaking up the turf on the high downs is often disastrous; the thin soil which was preserved by the close, hard turf is blown or washed away, and the soil becomes poorer year by year, in spite of dressing, until it is hardly worth cultivating. Clover may be grown on it, but it continues to deteriorate; or the tenant or landlord may turn it into a rabbit-warren, the most fatal policy of all. How hideous they are—those great stretches of downland, enclosed in big wire fences and rabbit netting, with little but wiry weeds, moss, and lichen growing on them, the earth dug up everywhere by the disorderly little beasts! For a while there is a profit—'it will serve me my time,' the owner says—but the end is utter barrenness.

One must lament, too, the destruction of the ancient earth-works, especially of the barrows, which is going on all over the downs, most rapidly where the land is broken up by the plough. One wonders if the ever-increasing curiosity of our day with regard to the history of the human race in the land continues to grow, what our descendants of the next half of the century, to go no further, will say of us and our incredible carelessness in the matter! So small a matter to us, but one which will, perhaps, be immensely important to them! It is, perhaps, better for our peace that we do not know; it would not be pleasant to have our children's and children's children's contemptuous expressions sounding in our prophetic ears. Perhaps we have no right to complain of the obliteration of these memorials of antiquity by the plough; the living are more than the dead, and in this case it may be said that we are only following the Artemisian example in consuming (in our daily bread) minute portions of the ashes of our old relations, albeit untearfully, with a cheerful countenance. Still one cannot but experience a shock on seeing the plough driven through an ancient, smooth turf, curiously marked with barrows, lynchetts, and other mysterious mounds and depressions, where sheep have been pastured for a thousand years, without obscuring these chance hieroglyphs scored by men on the surface of the hills.

It is not, however, only on the cultivated ground that the destruction is going on; the rabbit, too, is an active agent in demolishing the barrows and other earth-works. He burrows into the mound and throws out bushels of chalk and clay, which is soon washed down by the rains; he tunnels it through and through and sometimes makes it his village; then one day the farmer or keeper, who is not an archaeologist, comes along and puts his ferrets into the holes, and one of them, after drinking his fill of blood, falls asleep by the side of his victim, and the keeper sets to work with a pick and shovel to dig him out, and demolishes half the barrow to recover his vile little beast.

II

SALISBURY AS I SEE IT

The Salisbury of the villager—The cathedral from the meadows—Walks to Wilton and Old Sarum—The spire and a rainbow—Charm of Old Sarum—The devastation—Salisbury from Old Sarum—Leland's description—Salisbury and the village mind—Market-day—The infirmary—The cathedral —The lesson of a child's desire—In the streets again—An Apollo of the downs

To THE dwellers on the Plain Salisbury itself is an exceedingly important place—the most important in the world. For if they have seen a greater—London, let us say—it has left but a confused, a phantasmagoric image on the mind, an impression of endless thoroughfares and of innumerable people all apparently in a desperate hurry to do something, yet doing nothing; a labyrinth of streets and wilderness of houses, swarming with beings who have no definite object and no more to do with realities than so many lunatics, and are unconfined because they are so numerous that all the asylums in the world could not contain them. But of Salisbury they have a very clear image: inexpressibly rich as it is in sights, in wonders, full of people—hundreds of people in the streets and market-place—they can take it all in and know its meaning. Every man and woman, of all classes, in all that concourse, is there for some definite purpose which they can guess and understand; and the busy street and market, and red houses and soaring spire, are all one, and part and parcel too of their own lives in their own distant little village by the Avon or Wylye, or anywhere on the Plain. And that soaring spire which, rising so high above the red town, first catches the eye, the one object which gives unity and distinction to the whole picture is not more distinct in the mind than the entire Salisbury with its manifold interests and activities.

There is nothing in the architecture of England more beautiful than that same spire. I have seen it many times, far and near, from all points of view, and am never in or near the place but I go to some spot where I look at and enjoy the sight; but I will speak here of the two best points of view.

The nearest, which is the artist's favourite point, is from the meadows; there, from the waterside, you have the cathedral not too far away nor too near for a picture, whether on canvas or in the mind, standing amidst its great old trees, with nothing but the moist green meadows and the river between. One evening, during the late summer of this wettest season, when the rain was beginning to cease, I went out this way for my stroll, the pleasantest if not the only 'walk' there is in Salisbury. It is true, there are two others: one to Wilton by its long, shady avenue; the other to Old Sarum; but these are now motor-roads, and until the loathed hooting and dusting engines are thrust away into roads of their own there is little pleasure in them for the man on foot. The rain ceased, but the sky was still stormy, with a great blackness beyond the cathedral and still other black clouds coming up from the west behind me. Then the sun, near its setting, broke out, sending a flame of orange colour through the dark masses around it, and at the same time flinging a magnificent rainbow on that black cloud against which the immense spire stood wet with rain and flushed with light, so that it looked like a spire built of a stone impregnated with silver. Never had Nature so glorified man's work! It was indeed a marvellous thing to see, an effect so rare that in all the years I had known Salisbury, and the many times I had taken that stroll in all weathers, it was my first experience of such a thing. How lucky, then, was Constable to have seen it, when he set himself to paint his famous picture! And how brave he was and even wise to have attempted such a subject, one which, I am informed by artists with the brush, only a madman would undertake, however great a genius he might be. It was impossible, we know, even to a Constable, but we admire his failure nevertheless, even as we admire Turner's many failures; but when we go back to Nature we are only too glad to forget all about the picture.

The view from the meadows will not, in the future, I fear, seem so interesting to me; I shall miss the rainbow, and shall never see again except in that treasured image the great spire as Constable saw and tried to paint it. In like manner, though for a different reason, my future visits to Old Sarum will no longer give me the same pleasure experienced on former occasions.

Old Sarum stands over the Avon, a mile and a half from Salisbury; a round chalk hill about 300 feet high, in its round shape and isolation resembling a stupendous tumulus in which the giants of antiquity were buried, its steeply sloping, green sides ringed about with vast, concentric earth-works and ditches, the work of the 'old people,' as they say on the Plain, when referring to the ancient Britons, but how ancient, whether invading Celts or Aborigines—the true Britons, who possessed the land from neolithic times—even the anthropologists, the wise men of today, are unable to tell us. Later, it was a Roman station, one of the most important, and in after ages a great Norman castle and cathedral city, until early in the thirteenth century, when the old church was pulled down and a new and better one to last for ever was built in the green plain by many running waters. Church and people gone, the castle fell into ruin, though some believe it existed down to the fifteenth century; but from that time onwards the site has been a place of historical memories and a wilderness. Nature had made it a sweet and beautiful spot; the earth over the old buried ruins was covered with an elastic turf, jewelled with the bright little flowers of the chalk, the ramparts and ditches being all overgrown with a dense thicket of thorn, holly, elder, bramble, and ash, tangled up with ivy, briony, and traveller's-joy. Once only during the last five or six centuries some slight excavations were made when, in 1834, as the result of an excessively dry summer, the lines of the cathedral foundations were discernible on the surface. But it will no longer be the place it was, the Society of Antiquaries having received permission from the Dean and Chapter of Salisbury to work their sweet will on the site. That ancient, beautiful carcass, which had long made their mouths water, on which they have now fallen like a pack of hungry hyenas to tear off the old hide of green turf and burrow down to open to the light or drag out the deep, stony framework. The beautiful surrounding thickets, too, must go, they tell me, since you cannot turn the hill inside out without destroying the trees and bushes that crown it. What person who has known it and has often sought that spot for the sake of its ancient associations, and of the sweet solace they have found in the solitude, or for the noble view of the sacred city from its summit, will not deplore this fatal amiability of the authorities, this weak desire to please everyone and inability to say no to such a proposal!

But let me now return to the object which brings me to this spot; it was not to lament the loss of the beautiful, which cannot be preserved in our age—even this best one of all which Salisbury possessed cannot be preserved—but to look at Salisbury from this point of view. It is not as from 'the meadows' a view of the cathedral only, but of the whole town, amidst its circle of vast green downs. It has a beautiful aspect from that point: a red-brick and red-tiled town, set low on that circumscribed space, whose soft, brilliant green is in lovely contrast with the paler hue of the downs beyond, the perennial moist green of its water-meadows. For many swift, clear currents flow around and through Salisbury, and doubtless in former days there were many more channels in the town itself. Leland's description is worth quoting: 'There be many fair streates in the Cite Saresbyri, and especially the High Streate and Castle Streate. . . . Al the Streates in a maner, in New Saresbyri, hath little streamlettes and arms derivyd out of Avon that runneth through them. The site of the very town of Saresbyri and much ground thereabout is playne and low, and as a pan or receyvor of most part of the waters of Wiltshire.'

On this scene, this red town with the great spire, set down among water-meadows, encircled by paler green chalk hills, I look from the top of the inner and highest rampart or earth-work; or going a little distance down sit at ease on the turf to gaze at it by the hour. Nor could a sweeter resting-place be found, especially at the time of ripe elder-berries, when the thickets are purple with their clusters and the starlings come in flocks to feed on them, and feeding keep up a perpetual, low musical jangle about me.

It is not, however, of 'New Saresbyri' as seen by the tourist, with a mind full of history, archæology, and the æsthetic delight in cathedrals, that I desire to write, but of Salisbury as it appears to the dweller on the Plain. For Salisbury is the capital of the Plain, the head and heart of all those villages, too many to count, scattered far and wide over the surrounding country. It is the villager's own peculiar city, and even as the spot it stands upon is the 'pan or receyvor, of most part of the waters of Wiltshire,' so is it the receyvor of all he accomplishes in his laborious life, and thitherward flow all his thoughts and ambitions. Perhaps it is not so difficult for me as it would be for most persons who are not natives to identify myself with him and see it as he sees it. That greater place we have been in, that mighty, monstrous London, is ever present to the mind and is like a mist before the sight when we look at other places; but for me there is no such mist, no image so immense and persistent as to cover and obscure all others, and no such mental habit as that of regarding people as a mere crowd, a mass, a monstrous organism, in and on which each individual is but a cell, a scale. This feeling troubles and confuses my mind when I am in London, where we live 'too thick'; but quitting it I am absolutely free; it has not entered my soul and coloured me with its colour or shut me out from those who have never known it, even of the simplest dwellers on the soil who, to our sophisticated minds, may seem like beings of another species. This is my happiness—to feel, in all places, that I am one with them. To say, for instance, that I am going to Salisbury tomorrow, and catch the gleam in the children's eye and watch them, furtively watching me, whisper to one another that there will be something for them, too, on the morrow. To set out betimes and overtake the early carriers' carts on the road, each with its little cargo of packages and women with baskets and an old man or two, to recognize acquaintances among those who sit in front, and as I go on overtaking and passing carriers and the half-gipsy, little 'general dealer' in his dirty, ramshackle, little cart drawn by a rough, fast-trotting pony, all of us intent on business and pleasure, bound for Salisbury—the great market and emporium and place of all delights for all the great Plain. I remember that on my very last expedition, when I had come twelve miles in the rain and was standing at a street corner, wet to the skin, waiting for my carrier, a man in a hurry said to me, 'I say, just keep an eye on my cart for a minute or two while I run round to see somebody. I've got some fowls in it, and if you see anyone come poking round just ask them what they want—you can't trust every one. I'll be back in a minute.' And he was gone, and I was very pleased to watch his cart and fowls till he came back.

Business is business and must be attended to, in fair or foul weather, but for business with pleasure we prefer it fine on market-day. The one great and chief pleasure, in which all participate, is just to be there, to be in the crowd—a joyful occasion which gives a festive look to every face. The mere sight of it exhilarates like wine. The numbers—the people and the animals! The carriers' carts drawn up in rows on rows—carriers from a hundred little villages on the Bourne, the Avon, the Wylye, the Nadder, the Ebble, and from all over the Plain, each bringing its little contingent. Hundreds and hundreds more coming by train; you see them pouring down Fisherton Street in a continuous procession, all hurrying market-wards. And what a lively scene the market presents now, full of cattle and sheep and pigs and crowds of people standing round the shouting auctioneers! And horses, too, the beribboned hacks, and ponderous draught horses with manes and tails decorated with golden straw, thundering over the stone pavement as they are trotted up and down! And what a profusion of fruit and vegetables, fish and meat, and all kinds of provisions on the stalls, where women with baskets on their arms are jostling and bargaining! The Corn Exchange is like a huge beehive, humming with the noise of talk, full of brown-faced farmers in their riding and driving clothes and leggings, standing in knots or thrusting their hands into sacks of oats and barley. You would think that all the farmers from all the Plain were congregated there. There is a joyful contagion in it all. Even the depressed young lover, the forlornest of beings, repairs his wasted spirits and takes heart again. Why, if I've seen a girl with a pretty face today I've seen a hundred—and more. And *she* thinks they be so few she can treat me like that and barely give me a pleasant word in a month! Let her come to Salisbury and see how many there be!

And so with every one in that vast assemblage—vast to the dweller in the Plain. Each one is present as it were in two places, since each has in his or her heart the constant image of home—the little peaceful village in the remote valley; of father and mother and neighbours and children, in school just now, or at play, or home to dinner—home cares and concerns and the business in Salisbury. The selling and buying; friends and relations to visit or to meet in the market-place, and—how often!—the sick one to be seen at the Infirmary. This home of the injured and ailing, which is in the mind of so many of the people gathered together, is indeed the cord that draws and binds the city and the village closest together and makes the two like one.

That great, comely building of warm, red brick in Fisherton Street, set well back so that you can see it as a whole, behind its cedar and beech-trees—how familiar it is to the villagers! In numberless humble homes, in hundreds of villages of the Plain, and all over the surrounding country, the 'Infirmary' is a name of the deepest meaning, and a place of many sad and tender and beautiful associations. I heard it spoken of in a manner which surprised me at first, for I know some of the London poor and am accustomed to their attitude towards the metropolitan hospitals. The Londoner uses them very freely; they have come to be as necessary to him as the grocer's shop and the public-house, but for all the benefits he receives from them he has no faintest sense of gratitude, and it is my experience that if you speak to him of this he is roused to anger and demands, 'What are they for?' So far is he from having any thankful thoughts for all that has been given him for nothing and done for him and for his, if he has anything to say at all on the matter it is to find fault with the hospitals and cast blame on them for not having healed him more quickly or thoroughly.

This country town hospital and infirmary is differently regarded by the villagers of the Plain. It is curious to find how many among them are personally acquainted with it; perhaps it is not easy for anyone, even in this most healthy district, to get through life without sickness, and all are liable to accidents. The injured or afflicted youth, taken straight from his rough, hard life and poor cottage, wonders at the place he finds himself in—the wide, clean, airy room and white, easy bed, the care and skill of the doctors, the tender nursing by women, and comforts and luxuries, all without payment, but given as it seems to him out of pure divine love and compassion—all this comes to him as something strange, almost incredible. He suffers much perhaps, but can bear pain stoically and forget it when it is past, but the loving kindness he has experienced is remembered.

That is one of the very great things Salisbury has for the villagers, and there are many more which may not be spoken of, since we do not want to lose sight of the wood on account of the trees; only one must be mentioned for a special reason, and that is the cathedral. The villager is extremely familiar with it as he sees it from the market and the street and from a distance, from all the roads which lead him to Salisbury. Seeing it he sees everything beneath it—all the familiar places and objects, all the streets—High and Castle and Crane Streets, and many others, including Endless Street, which reminds one of Sydney Smith's last flicker of fun before that candle went out; and the 'White Hart' and the 'Angel' and 'Old George' and the humbler 'Goat' and 'Green Man' and 'Shoulder of Mutton,' with many besides; and the great red building with its cedar-tree, and the knot of men and boys standing on the bridge gazing down on the trout in the swift river below; and the market-place and its busy crowds—all the familiar sights and scenes that come under the spire like a flock of sheep on a burning day in summer, grouped about a great tree growing in the pasture-land. But he is not familiar with the interior of the great fane; it fails to draw him, doubtless because he has no time in his busy, practical life for the cultivation of the æsthetic faculties. There is a crust over that part of his mind; but it need not always and ever be so; the crust is not on the mind of the child.

Before a stall in the market-place a child is standing with her mother—a commonplace-looking, little girl of about twelve, blue-eyed, light-haired, with thin arms and legs, dressed, poorly enough, for her holiday. The mother, stoutish, in her best but much-worn black gown and a brown straw, out-of-shape hat, decorated with bits of ribbon and a few soiled and frayed artificial flowers. Probably she is the wife of a labourer who works hard to keep himself and family on fourteen shillings a week; and she, too, shows, in her hard hands and sunburnt face, with little wrinkles appearing, that she is a hard worker; but she is very jolly, for she is in Salisbury on market-day, in fine weather, with several shillings in her purse—a shilling for the fares, and perhaps eightpence for refreshments, and the rest to be expended in necessaries for the house. And now to increase the pleasure of the day she has unexpectedly run against a friend! There they stand, the two friends, basket on arm, right in the midst of the jostling crowd, talking in their loud, tinny voices at a tremendous rate; while the girl, with a half-eager, half-listless expression, stands by with her hand on her mother's dress, and every time there is a second's pause in the eager talk she gives a little tug at the gown and ejaculates 'Mother!' The woman impatiently shakes off the hand and says sharply, 'What now, Marty! Can't 'ee let me say just a word without bothering!' and on the talk runs again; then another tug and 'Mother!' and then, 'You promised, mother,' and by and by, 'Mother, you said you'd take me to the cathedral next time.'

Having heard so much I wanted to hear more, and addressing the woman I asked her why her child wanted to go. She answered me with a good-humoured laugh, ' 'Tis all because she heard 'em talking about it last winter, and she'd never been, and I says to her, "Never you mind, Marty, I'll take you there the next time I go to Salisbury." '

'And she's never forgot it,' said the other woman.

'Not she—Marty ain't one to forget.'

'And you been four times, mother,' put in the girl.

'Have I now! Well, 'tis too late now—half-past two, and we must be 't "Goat" at four.'

'Oh, mother, you promised!'

'Well, then, come along, you worriting child, and let's have it over or you'll give me no peace'; and away they went. And I would have followed to know the result if it had been in my power to look into that young brain and see the thoughts and feelings there as the crystal-gazer sees things in a crystal. In a vague way, with some very early

18

memories to help me, I can imagine it—the shock of pleased wonder at the sight of that immense interior, that far-extending nave with pillars that stand like the tall trunks of pines and beeches, and at the end the light screen which allows the eye to travel on through the rich choir, to see, with fresh wonder and delight, high up and far off, that glory of coloured glass as of a window half-open to an unimaginable place beyond—a heavenly cathedral to which all this is but a dim porch or passage!

We do not properly appreciate the educational value of such early experiences; and I use that dismal word not because it is perfectly right or for want of a better one, but because it is in everybody's mouth and understood by all. For all I know to the contrary, village schools may be bundled in and out of the cathedral from time to time, but that is not the right way, seeing that the child's mind is not the crowd-of-children's mind. But I can imagine that when we have a wiser, better system of education in the villages, in which books will not be everything, and to be shut up six or seven hours every day to prevent the children from learning the things that matter most—I can imagine at such a time that the schoolmaster or mistress will say to the village woman, 'I hear you are going to Salisbury tomorrow, or next Tuesday, and I want you to take Janie or little Dan or Peter, and leave him for an hour to play about on the cathedral green and watch the daws flying round the spire, and take a peep inside while you are doing your marketing.'

Back from the cathedral once more, from the infirmary, and from shops and refreshment-houses, out in the sun among the busy people, let us delay a little longer for the sake of our last scene.

It was past noon on a hot, brilliant day in August, and that splendid weather had brought in more people that I had ever before seen con-gregated in Salisbury, and never had the people seemed so talkative and merry and full of life as on that day. I was standing at a busy spot by a row of carriers' carts drawn up at the side of the pavement, just where there are three public-houses close together, when I caught sight of a young man of about twenty-two or twenty-three, a shep-herd in a grey suit and thick, iron-shod, old boots and brown leg-gings, with a soft, felt hat thrust jauntily on the back of his head, coming along towards me with that half-slouching, half-swinging gait peculiar to the men of the downs, especially when they are in the town on pleasure bent. Decidedly he was there on pleasure and had been indulging in a glass or two of beer (perhaps three) and was very happy, trolling out a song in a pleasant, musical voice as he swung along, taking no notice of the people stopping and turning round to stare after him, or of those of his own party who were following and trying to keep up with him, calling to him all the time to stop, to wait, to go slow, and give them a chance. There were seven following him: a stout, middle-aged woman, then a grey-haired old woman and two girls, and last a youngish, married woman with a small boy by the hand; and the stout woman, with a red, laughing face, cried out, 'Oh, Dave, do stop, can't 'ee! Where be going so fast, man—don't 'ee see we can't keep up with 'ee?' But he would not stop nor listen. It was his day out, his great day in Salisbury, a very rare occasion, and he was very happy. Then she would turn back to the others and cry, ' 'Tisn't no use, he won't bide for us—did 'ee ever see such a boy!' and laughing and perspiring she would start on after him again.

Now this incident would have been too trivial to relate had it not been for the appearance of the man himself—his powerful and perfect physique and marvellously handsome face—such a face as the old Greek sculptors have left to the world to be universally regarded and admired for all time as the most perfect. I do not think that this was my feeling only; I imagine that the others in that street who were standing still and staring after him had something of the same sense of surprise and admiration he excited in me. Just then it happened that there was a great commotion outside one of the public-houses, where a considerable party of gipsies in their little carts had drawn up, and were all engaged in a violent, confused altercation. Probably they, or one of them, had just disposed of a couple of stolen ducks, or a sheep-skin, or a few rabbits, and they were quarrelling over the division of the spoil. At all events they were violently excited, scowling at each other and one or two in a dancing rage, and had collected a crowd of amused lookers-on; but when the young man came singing by they all turned to stare at him.

As he came on I placed myself directly in his path and stared straight into his eyes—grey eyes and very beautiful; but he refused to see me; he stared through me like an animal when you try to catch its eyes, and went by still trolling out his song, with all the others streaming after him.

III
WINTERBOURNE BISHOP

A favourite village—Isolated situation—Appearance of the village—
Hedge-fruit—The winterbourne—Human interest—The home feeling—
Man in harmony with nature—Human bones thrown out by a rabbit
—A spot unspoiled and unchanged

OF THE few widely separated villages, hidden away among the lonely
downs in the large, blank spaces between the rivers, the one I love
best is Winterbourne Bishop. Yet of the entire number—I know them
all intimately—I daresay it would be pronounced by most persons the
least attractive. It has less shade from trees in summer and is more
exposed in winter to the bleak winds of this high country, from which-
ever quarter they may blow. Placed high itself on a wide, unwooded
valley or depression, with the low, sloping downs at some distance
away, the village is about as cold a place to pass a winter in as one
could find in this district. And, it may be added, the most inconvenient to live in at any time, the nearest town, or the easiest to get
to, being Salisbury, twelve miles distant by a hilly road. The only
means of getting to that great centre of life which the inhabitants possess is by the carrier's cart, which makes the weary four-hours'
journey once a week, on market-day. Naturally, not many of them see
that place of delights oftener than once a year, and some but once in
five or more years.

Then, as to the village itself, when you have got down into its one
long, rather winding street, or road. This has a green bank, five or six
feet high, on either side, on which stand the cottages, mostly facing
the road. Real houses there are none—buildings worthy of being called
houses in these great days—unless the three small farm-houses are considered better than cottages, and the rather mean-looking rectory—the
rector, poor man, is very poor. Just in the middle part, where the
church stands in its green churchyard, the shadiest spot in the village,
a few of the cottages are close together, almost touching, then farther
apart, twenty yards or so, then farther still, forty or fifty yards.
They are small, old cottages; a few have seventeenth-century dates cut
on stone tablets on their fronts, but the undated ones look equally old;

some thatched, others tiled, but none particularly attractive. Certainly they are without the added charm of a green drapery—creeper or ivy rose, clematis, and honeysuckle; and they are also mostly without the cottage-garden flowers, unprofitably gay like the blossoming furze, but dear to the soul; the flowers we find in so many of the villages along the rivers, especially in those of the Wylye valley to be described in a later chapter.

The trees, I have said, are few, though the churchyard is shady, where you can refresh yourself beneath its ancient beeches and its one wide-branching yew, or sit on a tomb in the sun when you wish for warmth and brightness. The trees growing by or near the street are mostly ash or beech, with a pine or two, old but not large; and there are small or dwarf yew-, holly-, and thorn-trees. Very little fruit is grown; two or three to half a dozen apple- and damson-trees are called an orchard, and one is sorry for the children. But in late summer and autumn they get their fruit from the hedges. These run up towards the downs on either side of the village, at right angles with its street; long, unkept hedges, beautiful with scarlet haws and traveller's-joy, rich in bramble and elder berries and purple sloes and nuts—a thousand times more nuts than the little dormice require for their own modest wants.

Finally, to go back to its disadvantages, the village is waterless; at all events in summer, when water is most wanted. Water is such a blessing and joy in a village—a joy for ever when it flows throughout the year, as at Nether Stowey and Winsford and Bourton-on-the-Water, to mention but three of all those happy villages in the land which are known to most of us! What man on coming to such places and watching the rushing, sparkling, foaming torrent by day and listening to its splashing, gurgling sounds by night, does not resolve that he will live in no village that has not a perennial stream in it! This unblessed, high and dry village has nothing but the winterbourne which gives it its name; a sort of surname common to a score or two of villages in Wiltshire, Dorset, Somerset, and Hants. Here the bed of the stream lies by the bank on one side of the village street, and when the autumn and early winter rains have fallen abundantly, the hidden reservoirs within the chalk hills are filled to overflowing; then the water finds its way out and fills the dry old channel and sometimes turns the whole street into a rushing river, to the immense joy of the village children. They are like ducks, hatched and reared at some upland farm where there was not even a muddy pool to dibble in. For a season (the wet one) the village women have water at their own doors and can go out and dip pails in it as often as they want. When spring comes it is still flowing merrily, trying to make you believe that it is going to flow for ever; beautiful, green water-loving plants and grasses spring up and flourish along the roadside, and you may see comfrey and water forget-me-not in flower. Pools, too, have been formed in some deep, hollow places; they are fringed with tall grasses, whitened over with bloom of water-crowfoot, and poa grass grows up from the bottom to spread its green tresses over the surface. Better still, by and by a couple of stray moorhens make their appearance in the pool—strange birds, coloured glossy olive-brown, slashed with white, with splendid scarlet and yellow beaks! If by some strange chance a shining blue kingfisher were to appear it could not create a greater excitement. So much attention do they receive that the poor strangers have no peace of their lives. It is a happy time for the children, and a good time for the busy housewife, who has all the water she wants for cooking and washing and cleaning—she may now dash as many pailfuls over her brick floors as she likes. Then the clear, swift current begins to diminish, and scarcely have you had time to notice the change than it is altogether gone! The women must go back to the well and let the bucket down, and laboriously turn and turn the handle of the windlass till it mounts to the top again. The pretty moist, green herbage, the graceful grasses, quickly wither away; dust and straws and rubbish from the road lie in the dry channel, and by and by it is filled with a summer growth of dock and loveless nettles which no child may touch with impunity.

No, I cannot think that any person for whom it had no association, no secret interest, would, after looking at this village with its dried-up winterbourne, care to make his home in it. And no person, I imagine, wants to see it; for it has no special attraction and is away from any road, at a distance from everywhere. I knew a great many villages in Salisbury Plain, and was always adding to their number, but there was no intention of visiting this one. Perhaps there is not a village on the Plain, or anywhere in Wiltshire for that matter, which sees fewer strangers. Then I fell in with the old shepherd whose life will be related in the succeeding chapter, and who, away from his native place, had no story about his past life and the lives of those he had known—no thought in his mind, I might almost say, which was not connected with the village of Winterbourne Bishop. And many of his anecdotes and reflections proved so interesting that I fell into the habit of putting them down in my notebook; until in the end the place itself, where he had followed his 'homely trade' so long, seeing and feeling so much, drew me to it. I knew there was 'nothing to see' in it, that it was without the usual attractions; that there was, in fact, nothing but the human interest, but that was enough. So I came to it to satisfy an idle curiosity—just to see how it would accord with the mental picture produced by his description of it. I came, I may say, prepared to like the place for the sole but sufficient reason that it had been his home. Had it not been for this feeling he had produced in me I should not, I imagine, have cared to stay long in it. As it was, I did stay, then came again and found that it was growing on me. I wondered why; for the mere interest in the old shepherd's life memories did not seem enough to account for this deepening attachment. It began to seem to me that I liked it more and more because of its very barrenness—the entire absence of all the features which make a place attractive, noble scenery, woods, and waters; deer parks and old houses, Tudor, Elizabethan, Jacobean, stately and beautiful, full of art treasures; ancient monuments and historical associations. There were none of these things; there was nothing here but that wide, vacant expanse, very thinly populated with humble, rural folk—farmers, shepherds, labourers—living in very humble houses. England is so full of riches in ancient monuments and grand and interesting and lovely buildings and objects and scenes, that it is perhaps too rich. For we may get into the habit of looking for such things, expecting them at every turn, every mile of the way.

I found it a relief, at Winterbourne Bishop, to be in a country which had nothing to draw a man out of a town. A wide, empty land, with nothing on it to look at but a furze-bush; or when I had gained the summit of the down, and to get a little higher still stood on the top of one of its many barrows, a sight of the distant village, its low, grey or reddish-brown cottages half hidden among its few trees, the square, stone tower of its little church looking at a distance no taller than a milestone. That emptiness seemed good for both mind and body: I could spend long hours idly sauntering or sitting or lying on the turf, thinking of nothing, or only of one thing—that it was a relief to have no thought about anything.

But no, something was secretly saying to me all the time, that it was more than what I have said which continued to draw me to this

24

vacant place—more than the mere relief experienced on coming back to nature and solitude, and the freedom of a wide earth and sky. I was not fully conscious of what the something more was until after repeated visits. On each occasion it was a pleasure to leave Salisbury behind and set out on that long, hilly road, and the feeling would keep

with me all the journey, even in bad weather, sultry or cold, or with the wind hard against me, blowing the white chalk dust into my eyes. From the time I left the turnpike to go the last two and a half to three miles by the side-road I would gaze eagerly ahead for a sight of my destination long before it could possibly be seen; until, on gaining the summit of a low, intervening down, the wished scene would be disclosed—the vale-like, wide depression, with its line of trees, blue-green in the distance, flecks of red and grey colour of the houses among them—and at that sight there would come a sense of elation, like that of coming home.

This in fact was the secret! This empty place was, in its aspect, despite the difference in configuration between down and undulating plain, more like the home of my early years than any other place known to me in the country. I can note many differences, but they do not deprive me of this home feeling; it is the likenesses that hold me, the spirit of the place, one which is not a desert with the desert's melancholy or sense of desolation, but inhabited, although thinly and by humble-minded men whose work and dwellings are unobtrusive. The final effect of this wide, green space with signs of human life and labour on it, and sight of animals—sheep and cattle—at various distances, is that we are not aliens here, intruders or invaders on the earth, living in it but apart, perhaps hating and spoiling it, but with the other animals are children of Nature, like them living and seeking our subsistence under her sky, familiar with her sun and wind and rain.

If some ostentatious person had come to this strangely quiet spot and raised a staring, big house, the sight of it in the landscape would have made it impossible to have such a feeling as I have described—this sense of man's harmony and oneness with nature. From how much of England has this expression which nature has for the spirit, which is so much more to us than beauty of scenery, been blotted out! This quiet spot in Wiltshire has been inhabited from of old, how far back in time the barrows raised by an ancient, barbarous people are there to tell us, and to show us how long it is possible for the race of men, in all stages of culture, to exist on the earth without spoiling it.

One afternoon when walking on Bishop Down I noticed at a distance of a hundred yards or more that a rabbit had started making a burrow in a new place and had thrown out a vast quantity of earth. Going to the spot to see what kind of chalk or soil he was digging so deeply in, I found that he had thrown out a human thigh-bone and a rib or two. They were of a reddish-white colour and had been embedded in a hard mixture of chalk and red earth. The following day I went again, and there were more bones, and every day after that the number increased until it seemed to me that he had brought out the entire skeleton, minus the skull, which I had been curious to see. Then the bones disappeared. The man who looked after the game had seen them, and recognizing that they were human remains had judiciously taken them away to destroy or stow them away in some safe place. For if the village constable had discovered them, or heard of their presence, he would perhaps have made a fuss and even thought it necessary to communicate with the coroner of the district. Such things occasionally happen, even in Wiltshire where the chalk hills are full of the bones of dead men, and a solemn Crowner's quest is held on the remains of a Saxon or Dane or an ancient Briton. When some important person—a Sir Richard Colt Hoare, for example, who dug up 379 barrows in Wiltshire, or a General Pitt Rivers—throws out human remains nobody minds, but if an unauthorized rabbit kicks out a lot of bones the matter should be inquired into.

But the man whose bones had been thus thrown out into the sunlight after lying so long at that spot, which commanded a view of the distant, little village looking so small in that immense, green space—who and what was he, and how long ago did he live on the earth—at Winterbourne Bishop, let us say? There were two barrows in that part of the down, but quite a stone's-throw away from the spot where the rabbit was working, so that he may not have been one of the people of that period. Still, it is probable that he was buried a very long time ago, centuries back, perhaps a thousand years, perhaps longer, and by chance there was a slope there which prevented the water from percolating, and the soil in which he had been desposited, under that close-knit turf which looked as if it had never been disturbed, was one in which bones might keep uncrumbled for ever.

The thought that occurred to me at the time was that if the man himself had come back to life after so long a period, to stand once more on that down surveying the scene, he would have noticed little change in it, certainly nothing of a startling description. The village itself, looking so small at that distance, in the centre of the vast depression, would probably not be strange to him. It was doubtless there as far back as history goes and probably still farther back in time. For at that point, just where the winterbourne gushes out from the low hills, is the spot man would naturally select to make his home. And he would see no mansion or big building, no puff of white steam and sight of a long, black train creeping over the earth, nor any other strange thing. It would appear to him even as he knew it before he fell asleep—the same familiar scene, with furze and bramble and bracken on the slope, the wide expanse with sheep and cattle grazing in the distance, and the dark green of trees in the hollows, and fold on fold of the low down beyond, stretching away to the dim, farthest horizon.

IV

A SHEPHERD OF THE DOWNS

Caleb Bawcombe—An old shepherd's love of his home—Fifty years
shepherding—Bawcombe's singular appearance—A tale of a titlark—Caleb
Bawcombe's father—Father and son—A grateful sportsman and Isaac
Bawcombe's pension—Death following death in old married couples—In a
village churchyard—A farm-labourer's gravestone and his story

IT IS now several years since I first met Caleb Bawcombe, a shepherd of the South Wiltshire Downs, but already old and infirm and past work. I met him at a distance from his native village, and it was only after I had known him a long time and had spent many afternoons and evenings in his company, listening to his anecdotes of his shepherding days, that I went to see his own old home for myself—the village of Winterbourne Bishop already described, to find it a place after my own heart. But as I have said, if I had never known Caleb and heard so much from him about his own life and the lives of many of his fellow-villagers, I should probably never have seen this village.

One of his memories was of an old shepherd named John, whose acquaintance he made when a very young man—John being at that time seventy-eight years old—on the Winterbourne Bishop farm, where he had served for an unbroken period of close on sixty years. Though so aged he was still head shepherd, and he continued to hold that place seven years longer—until his master, who had taken over old John with the place, finally gave up the farm and farming at the same time. He, too, was getting past work and wished to spend his declining years in his native village in an adjoining parish, where he owned some house and cottage property. And now what was to become of the old shepherd, since the new tenant had brought his own men with him?—and he, moreover, considered that John, at eighty-five, was too old to tend a flock on the hills, even of tegs. His old master, anxious to help him, tried to get him some employment in the village where he wished to stay; and failing in this, he at last offered him a cottage rent free in the village where he was going to live himself, and, in addition, twelve shillings a week for the rest of his life. It was in those days an exceedingly generous offer, but John refused it. 'Master.' he said, 'I be going to stay in my own native village, and if I can't make a living the parish'll have to keep I; but keep or not keep, here I be and here I be going to stay, where I was borned.'

From this position the stubborn old man refused to be moved, and there at Winterbourne Bishop his master had to leave him, although not without having first made him a sufficient provision.

The way in which my old friend, Caleb Bawcombe, told the story plainly revealed his own feeling in the matter. He understood and had the keenest sympathy with old John, dead now over half a century; or rather, let us say, resting very peacefully in that green spot under the old grey tower of Winterbourne Bishop church where as a small boy he had played among the old gravestones as far back in time as the middle of the eighteenth century. But old John had long survived wife and children, and having no one but himself to think of was at liberty to end his days where he pleased. Not so with Caleb, for, although his undying passion for home and his love of the shepherd's calling was as great as John's, he was not so free, and he was compelled at last to leave his native downs, which he may never see again, to settle for the remainder of his days in another part of the country.

Early in life he 'caught a chill' through long exposure to wet and cold in winter; this brought on rheumatic fever and a malady of the thigh, which finally affected the whole limb and made him lame for life. Thus handicapped he had continued as shepherd for close on fifty years, during which time his sons and daughters had grown up, married, and gone away, mostly to a considerable distance, leaving their aged parents alone once more. Then the wife, who was a strong woman and of an enterprising temper, found an opening for herself at a distance from home where she could start a little business. Caleb indignantly refused to give up shepherding in his place to take part in so unheard-of an adventure; but after a year or more of life in his lonely hut among the hills and cold, empty cottage in the village, he at length tore himself away from that beloved spot and set forth on the longest journey of his life—about forty-five miles—to join her and help in the work of her new home. Here a few years later I found him, aged seventy-two, but owing to his increasing infirmities looking considerably more. When he considered that his father, a shepherd before him on those same Wiltshire Downs, lived to eighty-six, and his mother to eighty-four, and that both were vigorous and led active lives almost to the end, he thought it strange that his own work should be so soon done. For in heart and mind he was still young; he did not want to rest yet.

Since that first meeting nine years have passed, and as he is actually better in health today than he was then, there is good reason to hope that his staying power will equal that of his father.

I was at first struck with the singularity of Caleb's appearance, and later by the expression of his eyes. A very tall, big-boned, lean, round-shouldered man, he was uncouth almost to the verge of grotesqueness, and walked painfully with the aid of a stick, dragging his shrunken and shortened bad leg. His head was long and narrow, and his high forehead, long nose, long chin, and long, coarse, grey whiskers, worn like a beard on his throat, produced a goat-like effect. This was heightened by the ears and eyes. The big ears stood out from his head, and owing to a peculiar bend or curl in the membrane at the top they looked at certain angles almost pointed. The hazel eyes were wonderfully clear, but that quality was less remarkable than the unhuman intelligence in them—fawn-like eyes that gazed steadily at you as one may gaze through the window, open back and front, of a house at the landscape beyond. This peculiarity was a little disconcerting at first, when, after making his acquaintance out of doors, I went in uninvited and sat down with him at his own fireside. The busy old wife talked of this and that, and hinted as politely as she knew how that I was in her way. To her practical, peasant mind there was no sense in my being there. 'He be a stranger to we, and we be strangers to he.' Caleb was silent, and his clear eyes showed neither annoyance nor pleasure but only their native, wild alertness, but the caste feeling is always less strong in the hill shepherd than in other men who are on the land; in some cases it will vanish at a touch, and it was so in this one. A canary in a cage hanging in the kitchen served to introduce the subject of birds captive and birds free. I said that I liked the little yellow bird, and was not vexed to see him in a cage, since he was cage-born; but I considered that those who caught wild birds and kept them prisoners did not properly understand things. This happened to be Caleb's view. He had a curiously tender feeling about the little wild birds, and one amusing incident of his boyhood which he remembered came out during our talk. He was out on the down one summer day in charge of his father's flock, when two boys of the village on a ramble in the hills came and sat down on the turf by his side. One of them had a titlark, or meadow pipit, which he had just caught, in his hand, and there was a hot argument as to which of the two was the lawful owner of the poor little captive. The facts were as follows. One of the boys having found the nest became possessed with the desire to get the bird. His companion at once offered to catch it for him, and together they withdrew to a distance and sat down and waited until the bird returned to sit on the eggs. Then the young birdcatcher returned to the spot, and creeping quietly up to within five or six feet of the nest threw his hat so that it fell over the sitting titlark; but after having thus secured it he refused to give it up. The dispute waxed hotter as they sat there, and at last when it got to the point of threats of cuffs on the ear and slaps on the face they agreed to fight it out, the victor to have the titlark. The bird was then put under a hat for safety on the smooth turf a few feet away, and the boys proceeded to take off their jackets and roll up their shirt-sleeves, after which they faced one another, and were just about to begin when Caleb, thrusting out his crook, turned the hat over and away flew the titlark.

The boys, deprived of their bird and of an excuse for a fight, would have discharged their fury on Caleb, but they durst not, seeing that his dog was lying at his side; they could only threaten and abuse him, call him bad names, and finally put on their coats and walk off.

That pretty little tale of a titlark was but the first of a long succession of memories of his early years, with half a century of shepherding life on the downs, which came out during our talks on many autumn and winter evenings as we sat by his kitchen fire. The earlier of these memories were always the best to me, because they took one back sixty years or more, to a time when there was more wildness in the earth than now, and a nobler wild animal life. Even more interesting were some of the memories of his father, Isaac Bawcombe, whose time went back to the early years of the nineteenth century. Caleb cherished an admiration and reverence for his father's memory which were almost a worship, and he loved to describe him as he appeared in his old age, when upwards of eighty. He was erect and tall, standing six feet two in height, well proportioned, with a clean-shaved, florid face, clear, dark eyes, and silver-white hair; and at this later period of his life he always wore the dress of an old order of pensioners to which he had been admitted—a soft, broad, white felt hat, thick boots and brown leather leggings, and a long, grey cloth overcoat with red collar and brass buttons.

According to Caleb, he must have been an exceedingly fine speci-

men of a man, both physically and morally. Born in 1800, he began
following a flock as a boy, and continued as shepherd on the same
farm until he was sixty, never rising to more than seven shillings a
week and nothing found, since he lived in the cottage where he was
born and which he inherited from his father. That a man of his fine
powers, a head-shepherd on a large hill-farm, should have had no
better pay than that down to the year 1860, after nearly half a century of work in one place, seems almost incredible. Even his sons, as
they grew up to man's estate, advised him to ask for an increase, but
he would not. Seven shillings a week he had always had; and that
small sum, with something his wife earned by making highly finished
smock-frocks, had been sufficient to keep them all in a decent way;
and his sons were now all earning their own living. But Caleb got married, and resolved to leave the old farm at Bishop to take a better
place at a distance from home, at Warminster, which had been
offered him. He would there have a cottage to live in, nine shillings a
week, and a sack of barley for his dog. At that time the shepherd had
to keep his own dog—no small expense to him when his wages were
no more than six to eight shillings a week. But Caleb was his father's
favourite son, and the old man could not endure the thought of losing
sight of him; and at last, finding that he could not persuade him not to
leave the old home, he became angry, and told him that if he went
away to Warminster for the sake of the higher wages and barley for
the dog he would disown him! This was a serious matter to Caleb, in
spite of the fact that a shepherd has no money to leave to his children
when he passes away. He went nevertheless, for, though he loved and
reverenced his father, he had a young wife who pulled the other way;
and he was absent for years, and when he returned the old man's
heart had softened, so that he was glad to welcome him back to the
old home.

Meanwhile at that humble cottage at Winterbourne Bishop great
things had happened; old Isaac was no longer shepherding on the
downs, but living very comfortably in his own cottage in the village.
The change came about in this way.

The downland shepherds, Caleb said, were as a rule clever poachers;
and it is really not surprising, when one considers the temptation to a
man with a wife and several hungry children, besides himself and a
dog, to feed out of about seven shillings a week. But old Bawcombe
was an exception: he would take no game, furred or feathered, nor, if

he could prevent it, allow another to take anything from the land fed by his flock. Caleb and his brothers, when as boys and youths they began their shepherding, sometimes caught a rabbit, or their dog caught and killed one without their encouragement; but, however the thing came into their hands, they could not take it home on account of their father. Now it happened that an elderly gentleman who had the shooting was a keen sportsman, and that in several successive years he found a wonderful difference in the amount of game at one spot among the hills and in all the rest of his hill property. The only explanation the keeper could give was that Isaac Bawcombe tended his flock on that down where rabbits, hares, and partridges were so plentiful. One autumn day the gentleman was shooting over that down, and seeing a big man in a smock-frock standing motionless, crook in hand, regarding him, he called out to his keeper, who was with him, 'Who is that big man?' and was told that it was Shepherd Bawcombe. The old gentleman pulled some money out of his pocket and said, 'Give him this half-crown, and thank him for the good sport I've had today.' But after the coin had been given the giver still remained standing there, thinking, perhaps, that he had not yet sufficiently rewarded the man; and at last, before turning away, he shouted, 'Bawcombe, that's not all. You'll get something more by and by.'

Isaac had not long to wait for the something more, and it turned out not to be the hare or brace of birds he had half expected. It happened that the sportsman was one of the trustees of an ancient charity which provided for six of the most deserving old men of the parish of Bishop; now, one of the six had recently died, and on this gentleman's recommendation Bawcombe had been elected to fill the vacant place. The letter from Salisbury informing him of his election and commanding his presence in that city filled him with astonishment; for, though he was sixty years old and the father of three sons now out in the world, he could not yet regard himself as an old man, for he had never known a day's illness, nor an ache, and was famed in all that neighbourhood for his great physical strength and endurance. And now, with his own cottage to live in, eight shillings a week, and his pensioners' garments, with certain other benefits, and a shilling a day besides which his old master paid him for some services at the farmhouse in the village, Isaac found himself very well off indeed, and he enjoyed his propserous state for twenty-six years. Then, in 1886, his old wife fell ill and died, and no sooner was she in her grave than he, too, began to droop; and soon, before the year was out, he followed her, because, as the neighbours said, they had always been a loving pair and one could not 'bide without the other.

This chapter has already had its proper ending and there was no intention of adding to it, but now for a special reason, which I trust the reader will pardon when he hears it, I must go on to say something about that strange phenomenon of death succeeding death in old married couples, one dying for no other reason than that the other has died. For it is our instinct to hold fast to life, and the older a man gets if he be sane the more he becomes like a newborn child in the impulse to grip tightly. A strange and a rare thing among people generally (the people we know), it is nevertheless quite common among persons of the labouring class in the rural districts. I have sometimes marvelled at the number of such cases to be met with in the villages; but when one comes to think about it one ceases to wonder that it should be so. For the labourer on the land goes on from boyhood to the end of life in the same everlasting round, the changes from task to task, according to the seasons, being no greater than in the case of the animals that alter their actions and habits to suit the varying conditions of the year. March and August and December, and every month, will bring about the changes in the atmosphere and earth and vegetation and in the animals, which have been from of old, which he knows how to meet, and the old, familiar task, lambing-time, shearing-time, root and seed crops, hoeing, haymaking, harvesting. It is a life of the extremest simplicity, without all those interests outside the home and the daily task, the innumerable distractions, common to all persons in other classes and to the workmen in towns as well. Incidentally it may be said that it is also the healthiest, that, speaking generally, the agricultural labourer is the healthiest and sanest man in the land, if not also the happiest, as some believe.

It is this life of simple, unchanging actions and of habits that are like instincts, of hard labour in sun and wind and rain from day to day, with its weekly break and rest, and of but few comforts and no luxuries, which serves to bind man and wife so closely. And the longer their life goes on together the closer and more unbreakable the union grows. They are growing old: old friends and companions have died or left them; their children have married and gone away and have their own families and affairs, so that the old folks at home are little remembered, and to all others they have become of little consequence in the world. But they do not know it, for they are together, cherishing the same memories, speaking of the same old, familiar things, and their lost friends and companions, their absent, perhaps estranged, children, are with them still in mind as in the old days. The past is with them more than the present, to give an undying interest to life; for they share it, and it is only when one goes, when the old wife gets the tea ready and goes mechanically to the door to gaze out, knowing that her tired man will come in no more to take his customary place and listen to all the things she has stored up in her mind during the day to tell him; and when the tired labourer comes in at dusk to find no old wife waiting to give him his tea and talk to him while he refreshes himself, he all at once realizes his position; he finds himself cut off from the entire world, from all of his kind. Where are they all? The enduring sympathy of that one soul that was with him till now had kept him in touch with life, had made it seem unchanged and unchangeable, and with that soul has vanished the old, sweet illusion as well as all ties, all common, human affection. He is desolate, indeed, alone in a desert world, and it is not strange that in many and many a case, even in that of a man still strong, untouched by disease and good for another decade or two, the loss, the awful solitude, has proved too much for him.

Such cases, I have said, are common, but they are not recorded, though it is possible with labour to pick them out in the church registers; but in the churchyards you do not find them, since the farm-labourer has only a green mound to mark the spot where he lies. Nevertheless, he is sometimes honoured with a gravestone, and last August I came by chance on one on which was recorded a case like that of Isaac Bawcombe and his life-mate.

The churchyard is in one of the prettiest and most secluded villages in the downland country described in this book. The church is ancient and beautiful and interesting in many ways, and the churchyard, too, is one of the most interesting I know, a beautiful, green, tree-shaded spot, with an extraordinary number of tombs and gravestones, many of them dated in the eighteenth and seventeenth centuries, inscribed with names of families which have long died out.

I went on that afternoon to pass an hour in the churchyard, and finding an old man in labourer's clothes resting on a tomb, I sat down and entered into conversation with him. He was seventy-nine, he told me, and past work, and he had three shillings a week from the parish; but he was very deaf and it fatigued me to talk to him, and seeing the church open I went in. On previous visits I had had a good deal of trouble to get the key, and to find it open now was a pleasant surprise. An old woman was there dusting the seats, and by and by, while I was talking with her, the old labourer came stumping in with his ponderous, iron-shod boots and without taking off his old, rusty hat, and began shouting at the church-cleaner about a pair of trousers he had given her to mend, which he wanted badly. Leaving them to their arguing I went out and began studying the inscriptions on the stones, so hard to make out in some instances; the old man followed and went his way; then the church-cleaner came out to where I was standing. 'A tiresome old man!' she said. 'He's that deaf he has to shout to hear himself speak, then you've got to shout back—and all about his old trousers!'

'I suppose he wants them,' I returned, 'and you promised to do them, so he has some reason for going at you about it.'

'Oh no, he hasn't,' she replied. 'The girl brought them for me to mend, and I said, "Leave them and I'll do them when I've time"—how did I know he wanted them in a hurry? A troublesome old man!'

By and by, taking a pair of spectacles out of her pocket, she put them on, and going down on her knees she began industriously picking the old, brown, dead moss out of the lettering on one side of the tomb. 'I'd like to know what it says on this stone,' she said.

'Well, you can read it for yourself, now you've got your glasses on.'

'I can't read. You see, I'm old—seventy-six years, and when I were little we were very poor and I couldn't get no schooling. I've got these glasses to do my sewing, and only put them on to get this stuff out so's you could read it. I'd like to hear you read it.'

I began to get interested in the old dame who talked to me so freely. She was small and weak-looking, and appeared very thin in her limp, old, faded gown; she had a meek, patient expression on her face, and her voice, too, like her face, expressed weariness and resignation.

'But if you have always lived here you must know what is said on this stone?'

'No, I don't; nobody never read it to me, and I couldn't read it because I wasn't taught to read. But I'd like to hear you read it.'

It was a long inscription to a person named Ash, gentleman, of this parish, who departed this life over a century ago, and was a man of a noble and generous disposition, good as a husband, a father, a friend, and charitable to the poor. Under all were some lines of verse, scarcely legible in spite of the trouble she had taken to remove the old moss from the letters.

She listened with profound interest, then said, 'I never heard all that before; I didn't know the name, though I've known this stone since I was a child. I used to climb on to it then. Can you read me another?'

I read her another and several more, then came to one which she said she knew—every word of it, for this was the grave of the sweetest, kindest woman that ever lived. Oh, how good this dear woman had been to her in her young married life more'n fifty years ago! If that dear lady had only lived it would not have been so hard for her when her trouble come!

'And what was your trouble?'

'It was the loss of my poor man. He was such a good man, a thatcher; and he fell from a rick and injured his spine, and he died, poor fellow, and left me with our five little children.' Then, having told me her own tragedy, to my surprise she brightened up and begged me to read other inscriptions to her.

I went on reading, and presently she said, 'No, that's wrong. There wasn't ever a Lampard in this parish. That I know.'

'You don't know! There certainly was a Lampard or it would not be stated here, cut in deep letters on this stone.'

'No, there wasn't a Lampard. I've never known such a name and I've lived here all my life.'

'But there were people living here before you came on the scene. He died a long time ago, this Lampard—in 1714, it says. And you are only seventy-six, you tell me; that is to say, you were born in 1835, and that would be one hundred and twenty-one years after he died.'

'That's a long time! It must be very old, this stone. And the church too. I've heard say it was once a Roman Catholic church. Is that true?'

'Why, of course it's true—all the old churches were, and we were all of that faith until a King of England had a quarrel with the Pope and determined he would be Pope himself as well as king in his own country. So he turned all the priests and monks out, and took their property and churches and had his own men put in. That was Henry VIII.'

'I've heard something about that king and his wives. But about Lampard, it do seem strange I've never heard that name before.'

'Not strange at all; it was a common name in this part of Wiltshire in former days; you find it in dozens of churchyards, but you'll find very few Lampards living in the villages. Why, I could tell you a dozen or twenty surnames, some queer, funny names, that were common in these parts not more than a century ago which seem to have quite died out.'

'I should like to hear some of them if you'll tell me.'

'Let me think a moment: there was Thorr, Pizzie, Gee, Every, Pottle, Kiddle, Toomer, Shergold, and—'

Here she interrupted to say that she knew three of the names I had mentioned. Then, pointing to a small, upright gravestone about twenty feet away, she added, 'And there's one.'

'Very well,' I said, 'but don't keep putting me out—I've got more names in my mind to tell you. Maidment, Marchmont, Velvin, Burpitt, Winzur, Rideout, Cullurne.'

Of these she only knew one—Rideout.

Then I went over to the stone she had pointed to and read the inscription to John Toomer and his wife Rebecca. She died first, in March 1877, aged 72: he in July the same year, aged 75.

'You knew them, I suppose?'

'Yes, they belonged here, both of them.'

'Tell me more about them.'

'There's nothing to tell; he was only a labourer and worked on the same farm all his life.'

'Who put a stone over them—their children?'

'No, they're all poor and live away. I think it was a lady who lived here; she'd been good to them, and she came and stood here when they put old John in the ground.'

'But I want to hear more.'

'There's no more, I've said; he was a labourer, and after she died he died.'

'Yes? go on.'

'How can I go on? There's no more. I knew them so well; they lived in the little thatched cottage over there, where the Millards live now.'

'Did they fall ill at the same time?'

'Oh no, he was as well as could be, still at work, till she died, then he went on in a strange way. He would come in of an evening and call his wife. "Mother! Mother, where are you?" you'd hear him call, "Mother, be you upstairs? Mother ain't you coming down for a bit of bread and cheese before you go to bed?" And then in a little while he just died.'

38

'And you said there was nothing to tell!'

'No, there wasn't anything. He was just one of us, a labourer on the farm.'

I then gave her something, and to my surprise after taking it she made me an elaborate curtsy. It rather upset me, for I had thought we had got on very well together and were quite free and easy in our talk, very much on a level. But she was not done with me yet. She followed to the gate, and holding out her open hand with small gift in it, she said in a pathetic voice, 'Did you think, sir, I was expecting this? I had no such thought and didn't want it.'

And I had no thought of saying or writing a word about her. But since that day she has haunted me—she and her old John Toomer, and it has just now occurred to me that by putting her in my book I may be able to get her out of my mind.

V

EARLY MEMORIES

A child shepherd—Isaac and his children—Shepherding in boyhood—Two notable sheep-dogs—Jack, the adder-killer—Sitting on an adder—Rough and the drovers—The Salisbury coach—A sheep-dog suckling a lamb

CALEB'S shepherding began in childhood; at all events he had his first experience of it at that time. Many an old shepherd, whose father was shepherd before him, has told me that he began to go with the flock very early in life, when he was no more than ten to twelve years of age. Caleb remembered being put in charge of his father's flock at the tender age of six. It was a new and wonderful experience, and made so vivid and lasting an impression on his mind that now, when he is past eighty, he speaks of it very feelingly as of something which happened yesterday.

It was harvesting time, and Isaac, who was a good reaper, was wanted in the field, but he could find no one, not even a boy, to take charge of his flock in the meantime, and so to be able to reap and keep an eye on the flock at the same time he brought his sheep down to the part of the down adjoining the field. It was on his 'liberty,' or that part of the down where he was entitled to have his flock. He then took his very small boy, Caleb, and placing him with the sheep told him they were now in his charge; that he was not to lose sight of them, and at the same time not to run about among the furze-bushes for fear of treading on an adder. By and by the sheep began straying off among the furze-bushes, and no sooner would they disappear from sight than he imagined they were lost for ever, or would be unless he quickly found them, and to find them he had to run about among the bushes with the terror of adders in his mind, and the two troubles together kept him crying with misery all the time. Then, at intervals, Isaac would leave his reaping and come to see how he was getting on, and the tears would vanish from his eyes, and he would feel very brave again, and to his father's question he would reply that he was getting on very well.

Finally his father came and took him to the field, to his great relief;

but he did not carry him in his arms; he strode along at his usual pace and let the little fellow run after him, stumbling and falling and picking himself up again and running on. And by and by one of the women in the field cried out, 'Be you not ashamed, Isaac, to go that pace and not bide for the little child! I do b'lieve he's no more'n seven year—poor mite!'

'No more'n six,' answered Isaac proudly, with a laugh.

But though not soft or tender with his children he was very fond of them, and when he came home early in the evening he would get them round him and talk to them, and sing old songs and ballads he had learnt in his young years—'Down in the Village,' 'The Days of Queen Elizabeth,' 'The Blacksmith,' 'The Gown of Green,' 'The Dawning of the Day,' and many others, which Caleb in the end got by heart and used to sing, too, when he was grown up.

Caleb was about nine when he began to help regularly with the flock; that was in the summer-time, when the flock was put every day on the down and when Isaac's services were required for the haymaking and later for harvesting and other work. His best memories of this period relate to his mother and to two sheepdogs, Jack at first and afterwards Rough, both animals of original character. Jack was a great favourite of his master, who considered him a 'tarrable good dog.' He was rather short-haired, like the old Welsh sheep-dog once common in Wiltshire, but entirely black instead of the usual colour—blue with a sprinkling of black spots. This dog had an intense hatred of adders and never failed to kill every one he discovered. At the same time he knew that they were dangerous enemies to tackle, and on catching sight of one his hair would instantly bristle up, and he would stand as if paralysed for some moments, glaring at it and gnashing his teeth, then springing like a cat upon it he would seize it in his mouth, only to hurl it from him to a distance. This action he would repeat until the adder was dead, and Isaac would then put it under a furze-bush to take it home and hang it on a certain gate. The farmer, too, like the dog, hated adders and paid his shepherd sixpence for every one his dog killed.

One day Caleb, with one of his brothers, was out with the flock, amusing themselves in their usual way on the turf with nine morris-men and the shepherd's puzzle, when all at once their mother appeared unexpectedly on the scene. It was her custom, when the boys were sent out with the flock, to make expeditions to the down just to see what they were up to; and hiding her approach by keeping to a hedge-side or by means of the furze-bushes, she would sometimes come upon them with disconcerting suddenness. On this occasion just where the boys had been playing there was a low, stout furze-bush, so dense and flat-topped that one could use it as a seat, and his mother taking off and folding her shawl placed it on the bush, and sat down

on it to rest herself after her long walk. 'I can see her now,' said Caleb, 'sitting on that furze-bush, in her smock and leggings, with a big hat like a man's on her head—for that's how she dressed.' But in a few moments she jumped up, crying out that she felt a snake under her, and snatched off the shawl, and there, sure enough out of the middle of the flat bush-top appeared the head of an adder, flicking out its tongue. The dog, too, saw it, dashed at the bush, forcing his muzzle and head into the middle of it, seized the serpent by its body and plucked it out and threw it from him, only to follow it up and kill it in the usual way.

Rough was a large, shaggy, grey-blue bobtail bitch with a white collar. She was a clever, good all-round dog, but had originally been trained for the road, and one of the shepherd's stories about her relates to her intelligence in her own special line—the driving of sheep.

One day he and his smaller brother were in charge of the flock on the down, and were on the side where it dips down to the turnpike-road about a mile and a half from the village, when a large flock, driven by two men and two dogs, came by. They were going to the Britford sheep-fair and were behind time; Isaac had started at daylight that morning with sheep for the same fair, and that was the reason of the boys being with the flock. As the flock on the down was feeding quietly the boys determined to go to the road to watch the sheep and men pass, and arriving at the roadside they saw that the dogs were too tired to work and the men were getting on with great difficulty. One of them, looking intently at Rough, asked if she would work. 'Oh, yes, she'll work,' said the boy proudly, and calling Rough he pointed to the flock moving very slowly along the road and over the turf on either side of it. Rough knew what was wanted; she had been looking on and had taken the situation in with her professional eye; away she dashed, and running up and down, first on one side then on the other, quickly put the whole flock, numbering 800 into the road and gave them a good start.

'Why, she be a road dog!' exclaimed the drover delightedly. 'She's better for me on the road than for you on the down; I'll buy her of you.'

'No, I mustn't sell her,' said Caleb.

'Look here, boy,' said the other, 'I'll give 'ee a sovran and this young dog, an' he'll be a good one with a little more training.'

'No, I mustn't,' said Caleb, distressed at the other's persistence.

'Well, will you come a little way on the road with us?' asked the drover.

This the boys agreed to and went on for about a quarter of a mile, when all at once the Salisbury coach appeared on the road, coming to meet them. This new trouble was pointed out to Rough, and at once when her little master had given the order she dashed barking into the midst of the mass of sheep and drove them furiously to the side from end to end of the extended flock, making a clear passage for the coach, which was not delayed a minute. And no sooner was the coach gone than the sheep were put back into the road.

Then the drover pulled out his sovereign once more and tried to make the boy take it.

'I mustn't,' he repeated, almost in tears. 'What would father say?'

'Say! He won't say nothing. He'll think you've done well.'

But Caleb thought that perhaps his father would say something, and when he remembered certain whippings he had experienced in the past he had an uncomfortable sensation about his back. 'No, I mustn't,' was all he could say, and then the drovers with a laugh went on with their sheep.

When Isaac came home and the adventure was told to him he laughed and said that he meant to sell Rough some day. He used to say this occasionally to tease his wife because of the dog's intense devotion to her; and she, being without a sense of humour and half thinking that he meant it, would get up out of her seat and solemnly declare that if he ever sold Rough she would never again go out to the down to see what the boys were up to.

One day she visited the boys when they had the flock near the turnpike, and seating herself on the turf a few yards from the road got out her work and began sewing. Presently they spied a big, singular-looking man coming at a swinging pace along the road. He was in shirt-sleeves, barefooted, and wore a straw hat without a rim. Rough eyed the strange being's approach with suspicion, and going to her mistress placed herself at her side. The man came up and sat down at a distance of three or four yards from the group, and Rough, looking dangerous, started up and put her forepaws on her mistress's lap and began uttering a low growl.

'Will that dog bite, missus?' said the man.

'Maybe he will,' said she. 'I won't answer for he if you come any nearer.'

The two boys had been occupied cutting a faggot from a furze-bush with a bill-hook, and now held a whispered consultation as to what they would do if the man tried to 'hurt mother,' and agreed that as soon as Rough had got her teeth in his leg they would attack him about the head with the bill-hook. They were not required to go into action; the stranger could not long endure Rough's savage aspect, and very soon he got up and resumed his travels.

The shepherd remembered another curious incident in Rough's career. At one time when she had a litter of pups at home she was yet compelled to be a great part of the day with the flock of ewes as they could not do without her. The boys just then were bringing up a motherless lamb by hand and they would put it with the sheep, and to feed it during the day were obliged to catch a ewe with milk. The lamb trotted at Caleb's heels like a dog and one day when it was hungry and crying to be fed, when Rough happened to be sitting on her haunches close by, it occurred to him that Rough's milk might serve as well as a sheep's. The lamb was put to her and took very kindly to its canine foster-mother, wriggling its tail and pushing vigorously with its nose. Rough submitted patiently to the trial, and the result was that the lamb adopted the sheep-dog as its mother and sucked her milk several times every day, to the great admiration of all who witnessed it.

44

VI

SHEPHERD ISAAC BAWCOMBE

A noble shepherd—A fighting village blacksmith—Old Joe the collier—A story of his strength—Donkeys poisoned by yew—The shepherd without his sheep—How the shepherd killed a deer

To me the most interesting of Caleb's old memories were those relating to his father, partly on account of the man's fine character, and partly because they went so far back, beginning in the early years of the last century.

Altogether he must have been a very fine specimen of a man, both physically and morally. In Caleb's mind he was undoubtedly the first among men morally, but there were two other men supposed to be his equals in bodily strength, one a native of the village, the other a periodical visitor. The first was Jarvis the blacksmith, a man of an immense chest and big arms, one of Isaac's greatest friends, and very good-tempered except when in his cups, for he did occasionally get drunk, and then he quarrelled with anyone and every one.

One afternoon he had made himself quite tipsy at the inn, and when going home, swaying about and walking all over the road, he all at once caught sight of the big shepherd coming soberly on behind. No sooner did he see him than it occurred to his wild and muddled mind that he had a quarrel with this very man, Shepherd Isaac, a quarrel of so pressing a nature that there was nothing to do but to fight it out there and then. He planted himself before the shepherd and challenged him to fight. Isaac smiled and said nothing.

'I'll fight thee about this,' he repeated, and began tugging at his coat, and after getting it off again made up to Isaac, who still smiled and said no word. Then he pulled his waistcoat off, and finally his shirt, and with nothing but his boots and breeches on once more squared up to Isaac and threw himself into his best fighting attitude.

'I doan't want to fight thee,' said Isaac at length, 'but I be thinking 'twould be best to take thee home.' And suddenly dashing in he seized Jarvis round the waist with one arm, grasped him round the legs with the other, and flung the big man across his shoulder, and carried him

45

off, struggling and shouting, to his cottage. There at the door, pale
and distressed, stood the poor wife waiting for her lord, when Isaac
arrived, and going straight in dropped the smith down on his own
floor, and with the remark, 'Here be your man,' walked off to his cottage and his tea.

The other powerful man was Old Joe the collier, who flourished and
was known in every village in the Salisbury Plain district during the
first thirty-five years of the last century. I first heard of this once
famous man from Caleb, whose boyish imagination had been affected
by his gigantic figure, mighty voice, and his wandering life over all
that wide world of Salisbury Plain. Afterwards when I became acquainted with a good many old men, aged from 75 to 90 and
upwards, I found that Old Joe's memory is still green in a good many
villages of the district, from the upper waters of the Avon to the
borders of Dorset. But it is only these ancients who knew him that
keep it green; by and by when they are gone Old Joe and his neddies
will be remembered no more.

In those days—down to about 1840 it was customary to burn peat
in the cottages, the first cost of which was about four and sixpence
the wagon-load—as much as I should require to keep me warm for a
month in winter; but the cost of its conveyance to the villages of the
Plain was about five to six shillings per load, as it came from a considerable distance, mostly from the New Forest. How the labourers at
that time, when they were paid seven or eight shillings a week, could
afford to buy fuel at such prices to bake their rye bread and keep the
frost out of their bones is a marvel to us. Isaac was a good deal better
off than most of the villagers in this respect, as his master—for he
never had but one—allowed him the use of a wagon and the driver's
services for the conveyance of one load of peat each year. The wagon-
load of peat and another of faggots lasted him the year with the furze
obtained from his 'liberty' on the down. Coal at that time was only
used by the blacksmiths in the villages, and was conveyed in sacks on
ponies or donkeys, and of those who were engaged in this business the
best known was Old Joe. He appeared periodically in the villages with
his eight donkeys, or neddies as he called them, with jingling bells on
their headstalls and their burdens of two sacks of small coal on each.
In stature he was a giant of about six feet three, very broad-chested,
and invariably wore a broad-brimmed hat, a slate-coloured smock-
frock, and blue worsted stockings to his knees. He walked behind the

donkeys, a very long staff in his hand, shouting at them from time to time, and occasionally swinging his long staff and bringing it down on the back of a donkey who was not keeping up the pace. In this way he wandered from village to village from end to end of the Plain, getting rid of his small coal and loading his animals with scrap iron which the blacksmiths would keep for him, and as he continued his rounds for nearly forty years he was a familiar figure to every inhabitant throughout the district.

There are some stories still told of his great strength, one of which is worth giving. He was a man of iron constitution and gave himself a hard life, and he was hard on his neddies, but he had to feed them well, and this he often contrived to do at some one else's expense. One night at a village on the Wylye it was discovered that he had put his eight donkeys in a meadow in which the grass was just ripe for mowing. The enraged farmer took them to the village pound and locked them up, but in the morning the donkeys and Joe with them had vanished and the whole village wondered how he had done it. The stone wall of the pound was four feet and a half high and the iron gate was locked, yet he had lifted the donkeys up and put them over and had loaded them and gone before anyone was up.

Once Joe met with a very great misfortune. He arrived late at a village, and finding there was good feed in the churchyard and that everybody was in bed, he put his donkeys in and stretched himself out among the gravestones to sleep. He had no nerves and no imagination; and was tired, and slept very soundly until it was light and time to put his neddies out before any person came by and discovered that he had been making free with the rector's grass. Glancing round he could see no donkeys, and only when he stood up he found they had not made their escape but were there all about him, lying among the gravestones, stone dead every one! He had forgotten that a churchyard was a dangerous place to put hungry animals in. They had browsed on the luxuriant yew that grew there, and this was the result.

In time he recovered from his loss and replaced his dead neddies with others, and continued for many years longer on his rounds.

To return to Isaac Bawcombe. He was born, we have seen, in 1800, and began following a flock as a boy and continued as shepherd on the same farm for a period of fifty-five years. The care of sheep was the one all-absorbing occupation of his life, and how much it was to him appears in this anecdote of his state of mind when he was deprived of it for a time. The flock was sold and Isaac was left without sheep, and with little to do except to wait from Michaelmas to Candlemas, when there would be sheep again at the farm. It was a long time to Isaac, and he found his enforced holiday so tedious that he made himself a nuisance to his wife in the house. Forty times a day he would throw off his hat and sit down, resolved to be happy at his own fireside, but after a few minutes the desire to be up and doing would return, and up he would get and out he would go again. One dark cloudy evening a man from the farm put his head in at the door. 'Isaac,' he said, 'there be sheep for 'ee up 't the farm—two hunderd ewes and a hunderd more to come in dree days. Master, he sent I to say you be wanted.' And away the man went.

Isaac jumped up and hurried forth without taking his crook from the corner and actually without putting on his hat! His wife called out after him, and getting no response sent the boy with his hat to over-take him. But the little fellow soon returned with the hat—he could not overtake his father!

He was away three or four hours at the farm, then returned, his hair very wet, his face beaming, and sat down with a great sigh of pleasure. 'Two hunderd ewes,' he said, 'and a hunderd more to come—what d'you think of that?'

'Well, Isaac,' said she, 'I hope thee'll be happy now and let I alone.'

After all that had been told to me about the elder Bawcombe's life and character, it came somewhat as a shock to learn that at one period during his early manhood he had indulged in one form of poaching—a sport which had a marvellous fascination for the people of England in former times, but was pretty well extinguished during the first quarter of the last century. Deer he had taken; and the whole tale of the deer-stealing, which was a common offence in that part of Wiltshire down to about 1834, sounds strange at the present day.

Large herds of deer were kept at that time at an estate a few miles from Winterbourne Bishop, and it often happened that many of the animals broke bounds and roamed singly and in small bands over the hills. When deer were observed in the open, certain of the villagers would settle on some plan of action; watchers would be sent out not only to keep an eye on the deer but on the keepers too. Much de-pended on the state of the weather and the moon, as some light was necessary; then, when the conditions were favourable and the keepers had been watched to their cottages, the gang would go out for a night's hunting. But it was a dangerous sport, as the keepers also knew that deer were out of bounds, and they would form some counter-plan, and one peculiarly nasty plan they had was to go out about three or four o'clock in the morning and secrete themselves somewhere close to the village to intercept the poachers on their return.

Bawcombe, who never in his life associated with the village idlers and frequenters of the alehouse, had no connexion with these men. His expeditions were made alone on some dark, unpromising night, when the regular poachers were in bed and asleep. He would steal away after bedtime, or would go out ostensibly to look after the sheep, and, if fortunate, would return in the small hours with a deer on his back. Then, helped by his mother, with whom he lived (for this was when he was a young unmarried man, about 1820), he would quickly skin and cut up the carcass, stow the meat away in some secret place, and bury the head, hide, and offal deep in the earth; and when morning came it would find Isaac out following his flock as usual, with no trace of guilt or fatigue in his rosy cheeks and clear, honest eyes.

This was a very astonishing story to hear from Caleb, but to suspect him of inventing or of exaggerating was impossible to anyone who knew him. And we have seen that Isaac Bawcombe was an exceptional man—physically a kind of Alexander Selkirk of the Wiltshire Downs. And he, moreover, had a dog to help him—one as superior in speed and strength to the ordinary sheep-dog as he himself was to the ruck of his fellow-men. It was only after much questioning on my part that Caleb brought himself to tell me of these ancient adventures, and finally to give a detailed account of how his father came to take his first deer. It was in the depth of winter—bitterly cold, with a strong north wind blowing on the snow-covered downs—when one evening Isaac caught sight of two deer out on his sheep-walk. In that part of Wiltshire there is a famous monument of antiquity, a vast mound-like wall, with a deep depression or fosse running at its side. Now it happened that on the highest part of the down, where the wall or mound was most exposed to the blast, the snow had been blown clean off the top, and the deer were feeding here on the short turf, keeping to the ridge, so that, outlined against the sky, they had become visible to Isaac at a great distance.

He saw and pondered. These deer, just now, while out of bounds, were no man's property, and it would be no sin to kill and eat one— if he could catch it!—and it was a season of bitter want. For many

many days he had eaten his barley bread, and on some days barley-flour dumplings, and had been content with this poor fare; but now the sight of these animals made him crave for meat with an intolerable craving, and he determined to do something to satisfy it.

He went home and had his poor supper, and when it was dark set forth again with his dog. He found the deer still feeding on the mound. Stealing softly along among the furze-bushes, he got the black line of the mound against the starry sky, and by and by, as he moved along, the black figures of the deer, with their heads down, came into view. He then doubled back and, proceeding some distance, got down into the fosse and stole forward to them again under the wall. His idea was that on taking alarm they would immediately make for the forest which was their home, and would probably pass near him. They did not hear him until he was within sixty yards, and then bounded down from the wall, over the dyke, and away, but in almost opposite directions—one alone making for the forest; and on this one the dog was set. Out he shot like an arrow from the bow, and after him ran Isaac 'as he had never runned afore in all his life.' For a short space deer and dog in hot pursuit were visible on the snow, then the darkness swallowed them up as they rushed down the slope; but in less than half a minute a sound came back to Isaac, flying, too, down the incline—the long, wailing cry of a deer in distress. The dog had seized his quarry by one of the front legs, a little above the hoof, and held it fast, and they were struggling on the snow when Isaac came up and flung himself upon his victim, then thrust his knife through its windpipe 'to stop its noise.' Having killed it, he threw it on his back and went home, not by the turnpike, nor by any road or path, but over fields and through copses until he got to the back of his mother's cottage. There was no door on that side, but there was a window and when he had rapped at it and his mother opened it, without speaking a word he thrust the dead deer through, then made his way round to the front.

That was how he killed his first deer. How the others were taken I do not know; I wish I did, since this one exploit of a Wiltshire shepherd has more interest for me than I find in fifty narratives of elephants slaughtered wholesale with explosive bullets, written for the delight and astonishment of the reading public by our most glorious Nimrods.

VII

THE DEER-STEALERS

Deer-stealing on Salisbury Plain—The head-keeper Harbutt—Strange story of a baby—Found as a surname—John Barter the village carpenter—How the keeper was fooled—A poaching attack planned—The fight—Head-keeper and carpenter—The carpenter hides his son—The arrest—Barter's sons forsake the village

THERE were other memories of deer-taking handed down to Caleb by his parents, and the one best worth preserving relates to the head-keeper of the preserves, or chase, and to a great fight in which he was engaged with two brothers of the girl who was afterwards to be Isaac's wife.

Here it may be necessary to explain that formerly the owner of Cranbourne Chase, at that time Lord Rivers, claimed the deer and the right to preserve and hunt deer over a considerable extent of country outside of his own lands. On the Wiltshire side these rights extended from Cranbourne Chase over the South Wiltshire Downs to Salisbury, and the whole territory, about thirty miles broad, was divided into beats or walks, six or eight in number, each beat provided with a keeper's lodge. This state of things continued to the year 1834, when the chase was 'disfranchised' by Act of Parliament.

The incident I am going to relate occurred about 1815 or perhaps two or three years later. The border of one of the deer walks was at a spot known as Three Downs Place, two miles and a half from Winterbourne Bishop. Here in a hollow of the downs there was an extensive wood, and just within the wood a large stone house, said to be centuries old but long pulled down, called Rollston House, in which the head-keeper lived with two under-keepers. He had a wife but no children, and was a middle-aged, thick-set, very dark man, powerful and vigilant, a 'tarrable' hater and persecutor of poachers, feared and hated by them in turn, and his name was Harbutt.

It happened that one morning, when he had unbarred the front door to go out, he found a great difficulty in opening it, caused by a heavy object having been fastened to the door-handle. It proved to be a basket or box, in which a well-nourished, nice-looking boy baby was

51

sleeping, well wrapped up and covered with a cloth. On the cloth a scrap of paper was pinned with the following lines written on it:

> Take me in and treat me well,
> For in this house my father dwell.

Harbutt read the lines and didn't even smile at the grammar; on the contrary, he appeared very much upset, and was still standing holding the paper, staring stupidly at it, when his wife came on the scene. 'What be this?' she exclaimed, and looked first at the paper, then at him, then at the rosy child fast asleep in its cradle; and instantly, with a great cry, she fell on it and snatched it up in her arms, and holding it clasped to her bosom, began lavishing caresses and endearing expressions on it, tears of rapture in her eyes! Not one word of inquiry or bitter, jealous repreach—all that part of her was swallowed up and annihilated in the joy of a woman who had been denied a child of her own to love and nourish and worship. And now one had come to her and it mattered little how. Two or three days later the infant was baptized at the village church with the quaint name of Moses Found.

Caleb was a little surprised at my thinking it a laughable name. It was to his mind a singularly appropriate one; he assured me it was not the only case he knew of in which the surname Found had been bestowed on a child of unknown parentage, and he told me the story of one of the Founds who had gone to Salisbury as a boy and worked and saved and eventually become quite a prosperous and important person. There was really nothing funny in it.

The story of Moses Found had been told him by his old mother; *she*, he remarked significantly, had good cause to remember it. She was herself a native of the village, born two or three years later than the mysterious Moses; her father, John Barter by name, was a carpenter and lived in an old, thatched house which still exists and is very familiar to me. He had five sons; then, after an interval of some years, a daughter was born, who in due time was to be Isaac's wife. When she was a little girl her brothers were all grown up or on the verge of manhood, and Moses, too, was a young man—'the spit of his father' people said, meaning the head-keeper—and he was now one of Harbutt's under-keepers.

About this time some of the more ardent spirits in the village, not satisfied with an occasional hunt when a deer broke out and roamed over the downs, took to poaching them in the woods. One night, a hunt having been arranged, one of the most daring of the men secreted himself close to the keeper's house, and having watched the keepers go in and the lights put out, he actually succeeded in fastening up the doors from the outside with screws and pieces of wood without creating an alarm. He then met his confederates at an agreed spot and the hunting began, during which one deer was chased to the house and actually pulled down and killed on the lawn.

Meanwhile the inmates were in a state of great excitement; the under-keepers feared that a force it would be dangerous to oppose had taken possession of the woods, while Harbutt raved and roared like a maddened wild beast in a cage, and put forth all his strength to pull the doors open. Finally he smashed a window and leaped out, gun in hand, and calling the others to follow rushed into the wood. But he was too late; the hunt was over and the poachers had made good their escape, taking the carcasses of two or three deer they had succeeded in killing.

The keeper was not to be fooled in the same way a second time, and before very long he had his revenge. A fresh raid was planned, and on this occasion two of the five brothers were in it, and there were four more, the blacksmith of Winterbourne Bishop, their best man, two famous shearers, father and son, from a neighbouring village, and a young farm labourer.

They knew very well that with the head-keeper in his present frame of mind it was a risky affair, and they made a solemn compact that if caught they would stand by one another to the end. And caught they were, and on this occasion the keepers were four.

At the very beginning the blacksmith, their ablest man and virtual leader, was knocked down senseless with a blow on his head with the butt end of a gun. Immediately on seeing this the two famous shearers took to their heels and the young labourer followed their example. The brothers were left but refused to be taken, although Harbutt roared at them in his bull's voice that he would shoot them unless they surrendered. They made light of his threats and fought against the four, and eventually were separated. By and by the younger of the two was driven into a brambly thicket where his opponents imagined that it would be impossible for him to escape. But he was a youth of indomitable spirit, strong and agile as a wild cat; and returning blow for blow he succeeded in tearing himself from them, then after a running fight through the darkest part of the wood for a distance of two or three hundred yards they at length lost him or gave him up and went back to assist Harbutt and Moses against the other man. Left to himself he got out of the wood and made his way back to the village. It was long past midnight when he turned up at his father's cottage, a pitiable object covered with mud and blood, hatless, his clothes torn to shreds, his face and whole body covered with bruises and bleeding wounds.

The old man was in a great state of distress about his other son, and early in the morning went to examine the ground where the fight had been. It was only too easily found; the sod was trampled down and branches broken as though a score of men had been engaged. Then he found his eldest son's cap, and a little farther away a sleeve of his coat; shreds and rags were numerous on the bramble bushes, and by and by he came on a pool of blood. 'They've kill 'n!' he cried in despair, 'they've killed my poor boy!' and straight to Rollston House he went to inquire, and was met by Harbutt himself, who came out limping, one boot on, the other foot bound up with rags, one arm in a sling and a cloth tied round his head. He was told that his son was alive and safe indoors and that he would be taken to Salisbury later in the day. 'His clothes be all torn to pieces,' added the keeper. 'You can just go home at once and git him others before the constable comes to take him.'

'You've tored them to pieces yourself and you can git him others,' retorted the old man in a rage.

'Very well,' said the keeper. 'But bide a moment—I've something more to say to you. When your son comes out of jail in a year or so you tell him from me that if he'll just step up this way I'll give him five shillings and as much beer as he likes to drink. I never see'd a better fighter!'

It was a great compliment to his son, but the old man was troubled in his mind. 'What dost mean, keeper, by a year or so?' he asked.

'When I said that,' returned the other, with a grin, 'I was just thinking what 'twould be he deserves to git.'

'And you'd agot your deserts, by God,' cried the angry father, 'if that boy of mine hadn't a-been left alone to fight ye!'

Harbutt regarded him with a smile of gratified malice. 'You can go home now,' he said. 'If you'd see your son you'll find'n in Salisbury jail. Maybe you'll be wanting new locks on your doors; you can git they in Salisbury too—you've no blacksmith in your village now. No, your boy weren't alone and you know that damned well.'

54

'I knew naught about that,' he returned, and started to walk home with a heavy heart. Until now he had been clinging to the hope that the other son had not been identified in the dark wood. And now what could he do to save one of the two from hateful imprisonment? The boy was not in a fit condition to make his escape; he could hardly get across the room and could not sit or lie down without groaning. He could only try to hide him in the cottage and pray that they would not discover him. The cottage was in the middle of the village and had but little ground to it, but there was a small, boarded-up cavity or cell at one end of an attic, and it might be possible to save him by putting him in there. Here, then, in a bed placed for him on the floor, his bruised son was obliged to lie, in the close, dark hole, for some days.

One day, about a week later, when he was recovering from his hurts, he crawled out of his box and climbed down the narrow stairs to the ground floor to see the light and breathe a better air for a short time, and while down he was tempted to take a peep at the street through the small, latticed window. But he quickly withdrew his head and by and by said to his father, 'I'm feared Moses has seen me. Just now when I was at the window he came by and looked up and see'd me with my head all tied up, and I'm feared he knew 'twas I.'

After that they could only wait in fear and trembling, and on the next day quite early there came a loud rap at the door, and on its being opened by the old man the constable and two keepers appeared standing before him.

'I've come to take your son,' said the constable.

The old man stepped back without a word and took down his gun from its place on the wall, then spoke:

'If you've got a search-warrant you may come in; if you haven't got 'n I'll blow the brains out of the first man that puts a foot inside my door.'

They hesitated a few moments then silently withdrew. After consulting together the constable went off to the nearest magistrate, leaving the two keepers to keep watch on the house: Moses Found was one of them. Later in the day the constable returned armed with a warrant and was thereupon admitted, with the result that the poor youth was soon discovered in his hiding-place and carried off. And that was the last he saw of his home, his young sister crying bitterly and his old father white and trembling with grief and impotent rage.

A month or two later the two brothers were tried and sentenced each to six months' imprisonment. They never came home. On their release they went to Woolwich, where men were wanted and the pay was good. And by and by the accounts they sent home induced first one then the other brother to go and join them, and the poor old father, who had been very proud of his five sons, was left alone with his young daughter—Isaac's destined wife.

VIII

SHEPHERDS AND POACHING

General remarks on poaching—Farmer, shepherd, and dog—A sheep-dog
that would not hunt—Taking a partridge from a hawk—Old Gaarge and
Young Gaarge—Partridge-poaching—The shepherd robbed of his rabbits—
Wisdom of Shepherd Gatherhood—Hare-trapping on the down—Hare-
taking with a crook

WHEN Caleb was at length free from his father's tutelage, and as an
under-shepherd practically independent, he did not follow Isaac's strict
example with regard to wild animals, good for the pot, which came
by chance in his way; he even allowed himself to go a little out of
his way on occasion to get them.

We know that about this matter the law of the land does not square
with the moral law as it is written in the heart of the peasant. A
wounded partridge or other bird which he finds in his walks abroad or
which comes by chance to him is his by a natural right, and he will
take and eat or dispose of it without scruple. With rabbits he is
very free—he doesn't wait to find a distressed one with a stoat on its
track—stoats are not sufficiently abundant; and a hare, too, may be
picked up at any moment; only in this case he must be very sure
that no one is looking. Knowing the law, and being perhaps a respectable, religious person, he is anxious to abstain from all appearance
of evil. This taking a hare or rabbit or wounded partridge is in his
mind a very different thing from systematic poaching; but he is aware
that to the classes above him it is not so—the law has made them
one. It is a hard, arbitrary, unnatural law, made by and for them, his
betters, and outwardly he must conform to it. Thus you will find the
best of men among the shepherds and labourers freely helping themselves to any wild creature that falls in their way, yet sharing the game-
preserver's hatred of the real poacher. The village poacher as a rule is
an idle, dissolute fellow, and the sober, industrious, righteous shepherd
or ploughman or carter does not like to be put on a level with such a
person. But there is no escape from the hard and fast rule in such
things, and however open and truthful he may be in everything else,
in this one matter he is obliged to practise a certain amount of decep-

tion. Here is a case to serve as an illustration; I have only just heard it, after putting together the material I had collected for this chapter, in conversation with an old shepherd friend of mine.

He is a fine old man who has followed a flock these fifty years, and will, I have no doubt, carry his crook for yet another ten. Not only is he a 'good shepherd,' in the sense in which Caleb uses that phrase, with a more intimate knowledge of sheep and all the ailments they are subject to than I have found in any other, but he is also a truly religious man, one that 'walks with God.' He told me this story of a sheep-dog he owned when head-shepherd on a large farm on the Dorsetshire border with a master whose chief delight in life was in coursing hares. They abounded on his land, and he naturally wanted the men employed on the farm to regard them as sacred animals. One day he came out to the shepherd to complain that some one had seen his dog hunting a hare.

The shepherd indignantly asked who had said such a thing.

'Never mind about that,' said the farmer. 'Is it true?'

'It is a lie,' said the shepherd. 'My dog never hunts a hare or anything else. 'Tis my belief the one that said that has got a dog himself that hunts the hares and he wants to put the blame on some one else.'

'May be so,' said the farmer, unconvinced.

Just then a hare made its appearance, coming across the field directly towards them, and either because they never moved or it did not smell them it came on and on, stopping at intervals to sit for a minute or so on its haunches, then on again until it was within forty yards of where they were standing. The farmer watched it approach and at the same time kept an eye on the dog sitting at their feet and watching the hare too, very steadily. 'Now shepherd,' said the farmer, 'don't you say one word to the dog and I'll see for myself.' Not a word did he say, and the hare came and sat for some seconds near them, then limped away out of sight and the dog made not the slightest movement. 'That's all right,' said the farmer, well pleased. 'I know now 'twas a lie I heard about your dog. I've seen for myself and I'll just keep a sharp eye on the man that told me.'

My comment on this story was that the farmer had displayed an almost incredible ignorance of a sheep-dog—and a shepherd. 'How would it have been if you had said, 'Catch him, Bob,' or whatever his name was?' I asked.

He looked at me with a twinkle in his eye and replied, 'I do b'lieve he'd ha' got 'n, but he'd never move till I told 'n.'

58

It comes to this: the shepherd refuses to believe that by taking a hare he is robbing any man of his property, and if he is obliged to tell a lie to save himself from the consequences he does not consider that it is a lie.

When he understood that I was on his side in this question, he told me about a good sheep-dog he once possessed which he had to get rid of because he would not take a hare!

A dog when broken is made to distinguish between the things he must and must not do. He is 'feelingly persuaded' by kind words and caresses in one case and hard words and hard blows in the other. He learns that if he hunts hares and rabbits it will be very bad for him, and in due time, after some suffering, he is able to overcome this strongest instinct of a dog. He acquires an artificial conscience. Then, when his education is finished, he must be made to understand that it is not quite finished after all—that he must partially unlearn one of the saddest of the lessons instilled in him. He must hunt a hare or rabbit when told by his master to do so. It is a compact between man and dog. Thus, they have got a law which the dog has sworn to obey; but the man who made it is above the law and can when he thinks proper command his servant to break it. The dog, as a rule, takes it all in very readily and often allows himself more liberty than his master gives him; the most highly accomplished animal is one that, like my shepherd's dog in the former instance, will not stir till he is told. In the other case the poor brute could not rise to the position; it was too complex for him, and when ordered to catch a rabbit he could only put his tail between his legs and look in a puzzled way at his master. 'Why do you tell me to do a thing for which I shall be thrashed?'

It was only after Caleb had known me some time, when we were fast friends, that he talked with perfect freedom of these things and told me of his own small, illicit takings without excuse or explanation.

One day he saw a sparrowhawk dash down upon a running partridge and struggle with it on the ground. It was in a grass field, divided from the one he was walking in by a large, unkept hedge without a gap in it to let him through. Presently the hawk rose up with the partridge still violently struggling in its talons, and flew over the hedge to Caleb's side, but was no sooner over than it came down again and the struggle went on once more on the ground. On Caleb running to the spot the hawk flew off, leaving his prey behind. He had grasped it in its sides, driving his sharp claws well in, and the partridge, though unable to fly, was still alive. The shepherd killed it and put it in his pocket, and enjoyed it very much when he came to eat it.

From this case, a most innocent form of poaching, he went on to relate how he had once been able to deprive a cunning poacher and bad man, a human sparrowhawk, of his quarry.

There were two persons in the village, father and son, he very heartily detested, known respectively as Old Gaarge and Young Gaarge, inveterate poachers both. They were worse than the real reprobate who haunted the public-house and did no work and was not ashamed of his evil ways, for these two were hypocrites and were outwardly sober, righteous men, who kept themselves a little apart from their neighbours and were very severe in their condemnation of other people's faults.

One Sunday morning Caleb was on his way to his ewes folded at a distance from the village, walking by a hedgerow at the foot of the down, when he heard a shot fired some way ahead, and after a minute or two a second shot. This greatly excited his curiosity and caused him to keep a sharp look-out in the direction the sounds had come from, and by and by he caught sight of a man walking towards him. It was Old Gaarge in his long smock-frock, proceeding in a leisurely way towards the village, but catching sight of the shepherd he turned aside through a gap in the hedge and went off in another direction to avoid meeting him. No doubt, thought Caleb, he has got his gun in two pieces hidden under his smock. He went on until he came to a small field of oats which had grown badly and had only been half reaped, and here he discovered that Old Gaarge had been lying in hiding to shoot at the partridges that came to feed. He had been screened from the sight of the birds by a couple of hurdles and some straw, and there were feathers of the birds he had shot scattered about. He had finished his Sunday morning's sport and was going back, a little too late on this occasion as it turned out.

Caleb went on to his flock, but before getting to it his dog discovered a dead partridge in the hedge; it had flown that far and then dropped, and there was fresh blood on its feathers. He put it in his pocket and carried it about most of the day while with his sheep on the down. Late in the afternoon he spied two magpies pecking at something out in the middle of a field and went to see what they had found. It was a second partridge which Old Gaarge had shot in the morning and had lost, the bird having flown to some distance before dropping. The magpies had probably found it already dead, as it was cold; they had begun tearing the skin at the neck and had opened it down to the breast-bone. Caleb took this bird, too, and by and by, sitting down to examine it, he thought he would try to mend the torn skin with the needle and thread he always carried inside his cap. He succeeded in stitching it neatly up, and putting back the feathers in their place the rent was quite concealed. That evening he took the two birds to a man in the village who made a livelihood by collecting bones, rags, and things of that kind; the man took the birds in his hand, held them up, felt their weight, examined them carefully, and pronounced them to be two good, fat birds, and agreed to pay two shillings for them.

Such a man may be found in most villages; he calls himself a 'general dealer,' and keeps a trap and pony—in some cases he keeps the ale-house—and is a useful member of the small, rural community—a sort of human carrion-crow.

The two shillings were very welcome, but more than the money was the pleasing thought that he had got the birds shot by the hypocritical old poacher for his own profit. Caleb had good cause to hate him. He, Caleb, was one of the shepherds who had his master's permission to take rabbits on the land, and having found his snares broken on many occasions he came to the conclusion that they were visited in the night time by some very cunning person who kept a watch on his movements. One evening he set five snares in a turnip field and went just before daylight next morning in a dense fog to visit them. Every one was broken! He had just started on his way back, feeling angry and much puzzled at such a thing, when the fog all at once passed away and revealed the figures of two men walking hurriedly off over the down. They were at a considerable distance, but the light was now strong enough to enable him to identify Old Gaarge and Young Gaarge. In a few moments they vanished over the brow. Caleb was mad at being deprived of his rabbits in this mean way, but pleased at the same time in having discovered who the culprits were; but what to do about it he did not know.

On the following day he was with his flock on the down and found himself near another shepherd, also with his sheep, one he knew very well, a quiet but knowing old man named Joseph Gathergood. He was known to be a skilful rabbit-catcher, and Caleb thought he would go over to him and tell him about how he was being tricked by the two Gaarges and ask him what to do in the matter.

The old man was very friendly and at once told him what to do. 'Don't you set no more snares by the hedges and in the turmots,' he said. 'Set them out on the open down where no one would go after rabbits and they'll not find the snares.' And this was how it had to be done. First he was to scrape the ground with the heel of his boot until the fresh earth could be seen through the broken turf; then he was to sprinkle a little rabbit scent on the scraped spot, and plant his snare. The scent and smell of the fresh earth combined would draw the rabbits to the spot; they would go there to scratch and would inevitably get caught if the snare was properly placed.

Caleb tried this plan with one snare, and on the following morning found that he had a rabbit. He set it again that evening, then again, until he had caught five rabbits on five consecutive nights, all with the same snare. That convinced him that he had been taught a valuable lesson and that old Gathergood was a very wise man about rabbits; and he was very happy to think that he had got the better of his two sneaking enemies.

But Shepherd Gathergood was just as wise about hares, and, as in the other case, he took them out on the down in the most open places. His success was due to his knowledge of the hare's taste for blackthorn twigs. He would take a good, strong blackthorn stem or shoot with twigs on it, and stick it firmly down in the middle of a large grass field or on the open down, and place the steel trap tied to the stick at a distance of a foot or so from it, the trap concealed under grass or moss and dead leaves. The smell of the blackthorn would draw the hare to the spot, and he would move round and round nibbling the twigs until caught.

Caleb never tried this plan, but was convinced that Gathergood was right about it.

He told me of another shepherd who was clever at taking hares in another way, and who was often chaffed by his acquaintances on account of the extraordinary length of his shepherd's crook. It was like a lance or pole, being twice the usual length. But he had a use for it. This shepherd used to make hares' forms on the downs in all suitable places, forming them so cunningly that no one seeing them by chance would have believed they were the work of human hands. The hares certainly made use of them. When out with his flock he would visit these forms, walking quietly past them at a distance of twenty to thirty feet, his dog following at his heels. On catching sight of a hare crouching in a form he would drop a word, and the dog would instantly stand still and remain fixed and motionless, while the shepherd went on but in a circle so as gradually to approach the form. Meanwhile the hare would keep his eyes fixed on the dog, paying no attention to the man, until by and by the long staff would be swung round and a blow descend on the poor, silly head from the opposite side, and if the blow was not powerful enough to stun or disable the hare, the dog would have it before it got many yards from the cosy nest prepared for its destruction.

IX

THE SHEPHERD ON FOXES

A fox-trapping shepherd—Gamekeepers and foxes—Fox and stoat—
A gamekeeper off his guard—Pheasants and foxes—Caleb kills a fox
—A fox-hunting sheep-dog—Two varieties of foxes—Rabbits playing
with little foxes—How to expel foxes—A playful spirit in the fox—
Fox-hunting a danger to sheep

CALEB related that his friend Shepherd Gathergood was a great fox-killer and, as with hares, he took them in a way of his own. He said that the fox will always go to a heap of ashes in any open place, and his plan was to place a steel trap concealed among the ashes, made fast to a stick about three feet high, firmly planted in the middle of the heap, with a piece of strong-smelling cheese tied to the top. The two attractions of an ash-heap and the smell of strong cheese was more than any fox could resist. When he caught a fox he killed and buried it on the down and said 'nothing to nobody' about it. He killed them to protect himself from their depredations; foxes, like Old Gaarge and his son in Caleb's case, went round at night to rob him of the rabbits he took in his snares.

Caleb never blamed him for this; on the contrary, he greatly admired him for his courage, seeing that if it had been found out he would have been a marked man. It was perhaps intelligence or cunning rather than courage; he did not believe that he would be found out, and he never was; he told Caleb of these things because he was sure of his man. Those who were interested in the hunt never suspected him, and as to gamekeepers, they hardly counted. He was helping them; no one hates a fox more than they do. The farmer gets compensation for damage, and the hen-wife is paid for her stolen chickens by the hunt. The keeper is required to look after the game, and at the same time to spare his chief enemy, the fox. Indeed, the keeper's state of mind with regard to foxes has always been a source of amusement to me, and by long practice I am able to talk to him on that delicate subject in a way to make him uncomfortable and self-contradictory. There are various, quite innocent questions which the student of wild life may put to a keeper about foxes which have a disturbing effect on his brain. How to expel foxes from a covert, for example; and here is another: Is it true that the fox listens for the distressed cries of a rabbit pursued by a stoat and that he will deprive the stoat of his captive? Perhaps; Yes; No, I don't think so, because one hunts by night, the other by day, he will answer, but you see that the question troubles him. One keeper, off his guard, promptly answered, 'I've no doubt of it; I can always bring a fox to me by imitating the cry of a rabbit hunted by a stoat.' But he did not say what his object was in attracting the fox.

I say that the keeper was off his guard in this instance, because the fiction that foxes were preserved on the estate was kept up, though as

a fact they were systematically destroyed by the keepers. As the pheasant-breeding craze appears to increase rather than diminish, notwithstanding the disastrous effect it has had in alienating the people from their lords and masters, the conflict of interest between fox-hunter and pheasant-breeder will tend to become more and more acute, and the probable end will be that fox-hunting will have to go. A melancholy outlook to those who love the country and old country sports, and who do not regard pheasant-shooting as now followed as sport at all. It is a delusion of the landlords that the country people think most highly of the great pheasant-preserver who has two or three big shoots in a season, during which vast numbers of birds are slaughtered—every bird 'costing a guinea,' as the saying is. It brings money into the country, he or his apologist tells you, and provides employment for the village poor in October and November, when there is little doing. He does not know the truth of the matter. A certain number of the poorer people of the village are employed as beaters for the big shoots at a shilling a day or so, and occasionally a labourer, going to or from his work, finds a pheasant's nest and informs the keeper and receives some slight reward. If he 'keeps his eyes open' and shows himself anxious at all times to serve the keeper he will sometimes get a rabbit for his Sunday dinner.

This is not a sufficient return for the freedom to walk on the land and in woods, which the villager possessed formerly, even in his worst days of his oppression, a liberty which has now been taken from him. The keeper is there now to prevent him; he was there before, and from of old, but the pheasant was not yet a sacred bird, and it didn't matter that a man walked on the turf or picked up a few fallen sticks in a wood. The keeper is there to tell him to keep to the road and sometimes to ask him, even when he is on the road, what is he looking over the hedge for. He slinks obediently away; he is only a poor labourer with his living to get, and he cannot afford to offend the man who stands between him and the lord and the lord's tenant. And he is inarticulate; but the insolence and injustice rankle in his heart, for he is not altogether a helot in soul; and the result is that the sedition-mongers, the Socialists, the furious denouncers of all landlords, who are now quartering the country, and whose vans I meet in the remotest villages, are listened to, and their words—wild and whirling words they may be—are sinking into the hearts of the agricultural labourers of the new generation.

To return to foxes and gamekeepers. There are other estates where the fiction of fox-preserving is kept up no longer, where it is notorious that the landlord is devoted exclusively to the gun and to pheasant-breeding. On one of the big estates I am familiar with in Wiltshire the keepers openly say they will not suffer a fox, and every villager knows it and will give information of a fox to the keepers, and looks to be rewarded with a rabbit. All this is undoubtedly known to the lord of the manor; his servants are only carrying out his own wishes, although he still subscribes to the hunt and occasionally attends the meet. The entire hunt may unite in cursing him, but they must do so below their breath; it would have a disastrous effect to spread it abroad that he is a persecutor of foxes.

66

Caleb disliked foxes, too, but not to the extent of killing them. He did once actually kill one, when a young under-shepherd, but it was accident rather than intention.

One day he found a small gap in a hedge, which had been made or was being used by a hare, and thinking to take it, he set a trap at the spot, tying it securely to a root and covering it over with dead leaves. On going to the place the next morning he could see nothing until his feet were on the very edge of the ditch, when with startling suddenness a big dog fox sprang up at him with a savage snarl. It was caught by a hind-leg, and had been lying concealed among the dead leaves close under the bank. Caleb, angered at finding a fox when he had looked for a hare, and at the attack the creature had made on him, dealt it a blow on the head with his heavy stick—just one blow given on the impulse of the moment, but it killed the fox! He felt very bad at what he had done and began to think of consequences. He took it from the trap and hid it away under the dead leaves beneath the hedge some yards from the gap, and then went to his work. During the day one of the farm hands went out to speak to him. He was a small, quiet old man, a discreet friend, and Caleb confided to him what he had done. 'Leave it to me,' said his old friend, and went back to the farm. In the afternoon Caleb was standing on the top of the down looking towards the village, when he spied at a great distance the old man coming out to the hills, and by and by he could make out that he had a sack on his back and a spade in his hand. When half-way up the side of the hill he put his burden down and set to work digging a deep pit. Into this he put the dead fox, and threw in and trod down the earth, then carefully put back the turf in its place, then, his task done, shouldered the spade and departed. Caleb felt greatly relieved, for now the fox was buried out on the downs, and no one would ever know that he had wickedly killed it.

Subsequently he had other foxes caught in traps set for hares, but was always able to release them. About one he had the following story. The dog he had at that time, named Monk, hated foxes as Jack hated adders, and would hunt them savagely whenever he got a chance. One morning Caleb visited a trap he had set in a gap in a hedge and found a fox in it. The fox jumped up, snarling and displaying his teeth, ready to fight for dear life, and it was hard to restrain Monk from flying at him. So excited was he that only when his master threatened him with his crook did he draw back and, sitting on his haunches, left him to deal with the difficult business in his own way. The difficulty was to open the steel trap without putting himself in the way of a bite from those 'tarrable sharp teeth.' After a good deal of manœuvring he managed to set the butt end of his crook on the handle of the gin, and forcing it down until the iron teeth relaxed their grip, the fox pulled his foot out, and darting away along the hedge side vanished into the adjoining copse. Away went Monk after him, in spite of his master's angry commands to him to come back, and fox and dog disappeared almost together among the trees. Sounds of yelping and of crashing through the undergrowth came back fainter and fainter, and then there was silence. Caleb waited at the spot full twenty minutes before the disobedient dog came back, looking very pleased. He had probably succeeded in overtaking and killing his enemy.

About that same Monk a sad story will have to be told in another chapter.

When speaking of foxes Caleb always maintained that in his part of the country there were two sorts: one small and very red, the larger one of a lighter colour with some grey in it. And it is possible that the hill foxes differed somewhat in size and colour from those of the lower country. He related that one year two vixens littered at one spot, a deep bottom among the downs, so near together that when the cubs were big enough to come out they mixed and played in company; the vixens happened to be of the different sorts, and the difference in colour appeared in the little ones as well.

Caleb was so taken with the pretty sight of all these little foxes, neighbours and playmates, that he went evening after evening to sit for an hour or longer watching them. One thing he witnessed which will perhaps be disbelieved by those who have not closely observed animals for themselves, and who still hold to the fable that all wild creatures are born with an inherited and instinctive knowledge and dread of their enemies. Rabbits swarmed at that spot, and he observed that when the old foxes were not about the young, half-grown rabbits would freely mix and play with the little foxes. He was so surprised at this, never having heard of such a thing, that he told his master of it, and the farmer went with him on a moonlight night and the two sat for a long time together, and saw rabbits and foxes playing, pursuing one another round and round, the rabbits when pursued often turning very suddenly and jumping clean over their pursuer.

The rabbits at this place belonged to the tenant, and the farmer, after enjoying the sight of the little ones playing together, determined to get rid of the foxes in the usual way by exploding a small quantity of gunpowder in the burrows. Four old foxes with nine cubs were too many for him to have. The powder was duly burned, and the very next day the foxes had vanished.

In Berkshire I once met with that rare being, an intelligent gamekeeper who took an interest in wild animals and knew from observation a great deal about their habits. During an after-supper talk, kept up till past midnight, we discussed the subject of strange, erratic actions in animals, which in some cases appear contrary to their own natures. He gave an instance of such behaviour in a fox that had its earth at a spot on the border of a wood where rabbits were abundant. One evening he was at this spot, standing among the trees and watching a number of rabbits feeding and gambolling on the green turf, when the fox came trotting by and the rabbits paid no attention. Suddenly he stopped and made a dart at a rabbit; the rabbit ran from him a distance of twenty to thirty yards, then suddenly turning round went for the fox and chased it back some distance, after which the fox again chased the rabbit, and so they went on, turn and turn about, half a dozen times. It was evident, he said, that the fox had no wish to catch and kill a rabbit, that it was nothing but play on his part, and that the rabbits responded in the same spirit, knowing that there was nothing to fear.

Another instance of this playful spirit of the fox with an enemy, which I heard recently, is of a gentleman who was out with his dog, a fox-terrier, for an evening walk in some woods near his house. On his way back he discovered on coming out of the woods that a fox was following him, at a distance of about forty yards. When he stood still the fox sat down and watched the dog. The dog appeared indifferent to its presence until his master ordered him to go for the fox, whereupon he charged him and drove him back to the edge of the wood, but at that point the fox turned and chased the dog right back to its master, then once more sat down and appeared very much at his ease. Again the dog was encouraged to go for him and hunted him again back to the wood, and was then in turn chased back to its master. After several repetitions of this performance, the gentleman went home, the fox still following, and on going in closed the gate behind him, leaving the fox outside, sitting in the road as if waiting for him to come out again to have some more fun.

This incident serves to remind me of an experience I had one evening in King's Copse, an immense wood of oak and pine in the New Forest near Exbury. It was growing dark when I heard on or close to the ground, some twenty to thirty yards before me, a low, wailing cry, resembling the hunger-cry of the young, long-eared owl. I began cautiously advancing, trying to see it, but as I advanced the cry receded, as if the bird was flitting from me. Now, just after I had begun following the sound, a fox uttered his sudden, startlingly loud scream about forty yards away on my right hand, and the next moment a second fox screamed on my left, and from that time I was accompanied, or shadowed, by the two foxes, always keeping abreast of me, always at the same distance, one screaming and the other replying about every half-minute. The distressful bird-sound ceased, and I turned and went off in another direction, to get out of the wood on the side nearest the place where I was staying, the foxes keeping with me until I was out.

What moved them to act in such a way is a mystery, but it was perhaps play to them.

Another curious instance of foxes playing was related to me by a gentleman at the little village of Inkpen, near the Beacon, in Berkshire. He told me that when it happened, a good many years ago, he sent an account of it to the 'Field.' His gamekeeper took him one day 'to see a strange thing,' to a spot in the woods where a fox had a litter of four cubs, near a long, smooth, green slope. A little distance from the edge of the slope three round swedes were lying on the turf. 'How do you think these swedes came here?' said the keeper, and then proceeded to say that the old fox must have brought them there from the field a long distance away, for her cubs to play with. He had watched them of an evening, and wanted his master to come and see too. Accordingly they went in the evening, and hiding themselves among the bushes near waited till the young foxes came out and began rolling the swedes about and jumping at and tumbling over them. By and by one rolled down the slope, and the young foxes went after it all the way down, and then, when they had worried it sufficiently, they returned to the top and played with another swede until that was rolled down, then with the third one in the same way. Every morning, the keeper said, the swedes were found back on top of the ground, and he had no doubt that they were taken up by the old fox again and left there for her cubs to play with.

Caleb was not so eager after rabbits as Shepherd Gathergood, but he disliked the fox for another reason. He considered that the hunted fox was a great danger to sheep when the ewes were heavy with lambs and when the chase brought the animal near if not right into the flock. He had one dreadful memory of a hunted fox trying to lose itself in his flock of heavy-sided ewes and the hounds following it and driving the poor sheep mad with terror. The result was that a large number of lambs were cast before their time and many others were poor, sickly things; many of the sheep also suffered in health. He had no extra money from the lambs that year. He received but a shilling (half a crown is often paid now) for every lamb above the number of ewes, and as a rule received from three to six pounds a year from this source.

X

BIRD LIFE ON THE DOWNS

Great bustard—Stone curlew—Big hawks—Former abundance of the raven—Dogs fed on carrion—Ravens fighting—Ravens' breeding-places in Wilts—Great Ridge Wood ravens—Field-fare breeding in Wilts—Pewit—Mistle-thrush—Magpie and turtledove—Gamekeepers and magpies—Rooks and farmers—Starling, the shepherd's favourite bird—Sparrowhawk and 'brown thrush'

WILTSHIRE, like other places in England, has long been deprived of its most interesting birds—the species that were best worth preserving. Its great bustard, once our greatest bird—even greater than the golden and sea eagles and the 'giant crane' with its 'trumpet sound' once heard in the land—is now but a memory. Or a place name: Bustard Inn, no longer an inn, is well known to the many thousands who now go to the mimic wars on Salisbury Plain; and there is a Trappist monastery in a village on the southernmost border of the county, which was once called, and is still known to old men, as 'Bustard Farm.' All that Caleb Bawcombe knew of this grandest bird is what his father had told him; and Isaac knew of it only from hearsay, although it was still met with in South Wilts when he was a young man.

The stone curlew, our little bustard with the long wings, big, yellow eyes, and wild voice, still frequents the uncultivated downs, unhappily in diminishing numbers. For the private collector's desire to possess British-taken birds' eggs does not diminish; I doubt if more than one clutch in ten escapes the searching eyes of the poor shepherds and labourers who are hired to supply the cabinets. One pair haunted a flinty spot at Winterbourne Bishop until a year or two ago; at other points a few miles away I watched other pairs during the summer of 1909, but in every instance their eggs were taken.

The larger hawks and the raven, which bred in all the woods and forests of Wiltshire, have, of course, been extirpated by the game-keepers. The biggest forest in the county now affords no refuge to any hawk above the size of a kestrel. Savernake is extensive enough, one would imagine, for condors to hide in, but it is not so. A few years ago a buzzard made its appearance there—just a common buzzard, and

72

the entire surrounding population went mad with excitement about it, and every man who possessed a gun flew to the forest to join in the hunt until the wretched bird, after being blazed at for two or three days, was brought down.

I heard of another case at Fonthill Abbey. Nobody could say what this wandering hawk was—it was very big, blue above with a white breast barred with black—a 'tarrable,' fierce-looking bird with fierce, yellow eyes. All the gamekeepers and several other men with guns were in hot pursuit of it for several days, until some one fatally wounded it, but it could not be found where it was supposed to have fallen. A fortnight later its carcass was discovered by an old shepherd, who told me the story. It was not in a fit state to be preserved, but he described it to me, and I have no doubt that it was a goshawk.

The raven survived longer, and the Shepherd Bawcombe talks about its abundance when he was a boy, seventy or more years ago. His way of accounting for its numbers at that time and its subsequent, somewhat rapid disappearance greatly interested me.

We have seen his account of deer-stealing by the villagers in those brave, old, starvation days when Lord Rivers owned the deer and hunting rights over a large part of Wiltshire, extending from Cranborne Chase to Salisbury, and when even so righteous a man as Isaac Bawcombe was tempted by hunger to take an occasional deer, discovered out of bounds. At that time, Caleb said, a good many dogs used for hunting the deer were kept a few miles from Winterbourne Bishop and were fed by the keepers in a very primitive manner. Old, worn-out horses were bought and slaughtered for the dogs. A horse would be killed and stripped of its hide somewhere away in the woods, and left for the hounds to batten on its flesh, tearing at and fighting over it like so many jackals. When only partially consumed the carcass would become putrid; then another horse would be killed and skinned at another spot perhaps a mile away, and the pack would start feeding afresh there. The result of so much carrion lying about was that ravens were attracted in numbers to the place and were so numerous as to be seen in scores together. Later, when the deer-hunting sport declined in the neighbourhood, and dogs were no longer fed on carrion, the birds decreased year by year, and when Caleb was a boy of nine or ten their former great abundance was but a memory. But he remembers that they were still fairly common, and he had much to say about the old belief that the raven 'smells death,' and when seen hovering over a flock, uttering its croak, it is a sure sign that a sheep is in a bad way and will shortly die.

One of his recollections of the bird may be given here. It was one of those things seen in boyhood which had very deeply impressed him. One fine day he was on the down with an elder brother, when they heard the familiar croak and spied three birds at a distance engaged in a fight in the air. Two of the birds were in pursuit of the third, and rose alternately to rush upon and strike at their victim from above. They were coming down from a considerable height, and at last were directly over the boys, not more than forty or fifty feet from the ground; and the youngsters were amazed at their fury, the loud, rushing sound of their wings, as of a torrent, and of their deep, hoarse croaks and savage, barking cries. Then they began to rise again, the hunted bird trying to keep above his enemies, they in their turn striving to rise higher still so as to rush down upon him from overhead; and in this way they towered higher and higher, their barking cries coming fainter and fainter back to earth, until the boys, not to lose sight of them, cast themselves down flat on their backs, and, continuing to gaze up, saw them at last no bigger than three 'leetle blackbirds.' Then they vanished, but the boys, still lying on their backs, kept their eyes on the same spot, and by and by first one black speck reappeared, then a second, and they soon saw that two birds were swiftly coming down to earth. They fell swiftly and silently, and finally pitched upon the down not more than a couple of hundred yards from the boys. The hunted bird had evidently succeeded in throwing them off and escaping. Probably it was one of their own young, for the ravens' habit is when their young are fully grown to hunt them out of the neighbourhood, or, when they cannot drive them off, to kill them.

There is no doubt that the carrion did attract ravens in numbers to this part of Wiltshire, but it is a fact that up to that date—about 1830—the bird had many well-known, old breeding-places in the county. The Rev. A. C. Smith, in his 'Birds of Wiltshire,' names twenty-three breeding-places, no fewer than nine of them on Salisbury Plain; but at the date of the publication of his work, 1887, only three of all these nesting-places were still in use: South Tidworth, Wilton Park, and Compton Park, Compton Chamberlain. Doubtless there were other ancient breeding-places which the author had not heard of: one was at the Great Ridge Wood, overlooking the Wylye valley, where ravens bred down to about thirty-five or forty years ago. I have found many old men in that neighbourhood who remember the birds, and they tell that the raven tree was a great oak which was cut down about sixty years ago, after which the birds built their nest in another tree not far away. A London friend of mine, who was born in the neighbourhood of the Great Ridge Wood, remembers the ravens as one of the common sights of the place when he was a boy. He tells of an unlucky farmer in those parts whose sheep fell sick and died in numbers, year after year, bringing him down to the brink of ruin, and how his old head-shepherd would say, solemnly shaking his head, ' 'Tis not strange—master, he shot a raven.'

There was no ravens' breeding-place very near Winterbourne Bishop. Caleb had 'never heared tell of a nestie'; but he had once seen the nest of another species which is supposed never to breed in this country. He was a small boy at the time, when one day an old shepherd of the place going out from the village saw Caleb, and calling to him said, 'You're the boy that likes birds; if you'll come with me, I'll show 'ee what no man ever seed afore'; and Caleb, fired with curiosity, followed him away to a distance from home, out from the downs, into the woods and to a place where he had never been, where there was bracken and heath with birch- and thorn-trees scattered about. On cautiously approaching a clump of birches they saw a big, thrush-like bird fly out of a large nest about ten feet from the ground, and settle on a tree close by, where it was joined by its mate. The old man pointed out that it was a felt or fieldfare, a thrush nearly as big as the mistle-thrush but different in colour, and he said that it was a bird that came to England in flocks in winter from no man knows where, far off in the north, and always went away before breeding-time. This was the only felt he had ever seen breeding in this country, and he 'didn't believe that no man had ever seed such a thing before.' He would not climb the tree to see the eggs, or even go very near it, for fear of disturbing the birds.

This man, Caleb said, was a great one for birds: he knew them all, but seldom said anything about them; he watched and found out a good deal about them just for his private pleasure.

The characteristic species of this part of the down country, comprising the parish of Winterbourne Bishop, are the pewit, magpie, turtle-dove, mistle-thrush, and starling. The pewit is universal on the hills, but will inevitably be driven away from all that portion of Salisbury

Plain used for military purposes. The mistle-thrush becomes common in summer after its early breeding season is ended, when the birds in small flocks resort to the downs, where they continue until cold weather drives them away to the shelter of the wooded, low country.

In this neighbourhood there are thickets of thorn, holly, bramble, and birch growing over hundreds of acres of down, and here the hill-magpie, as it is called, has its chief breeding-ground, and is so common that you can always get a sight of at least twenty birds in an afternoon's walk. Here, too, is the metropolis of the turtledove, and the low sound of its crooning is heard all day in summer, the other most common sound being that of magpies—their subdued, conversational chatter and their solo-singing, the chant or call which a bird will go on repeating for a hundred times. The wonder is how the doves succeed in such a place in hatching any couple of chalk-white eggs, placed on a small platform of sticks, or of rearing any pair of young, conspicious in their blue skins and bright yellow down!

The keepers tell me they get even with these kill-birds later in the year, when they take to roosting in the woods, a mile away in the valley. The birds are waited for at some point where they are accustomed to slip in at dark, and one keeper told me that on one evening alone assisted by a friend he had succeeded in shooting thirty birds.

On Winterbourne Bishop Down and round the village the magpies are not persecuted, probably because the gamekeepers, the professional bird-killers, have lost heart in this place. It is a curious and rather pretty story. There is no squire, as we have seen; the farmers have the rabbits and for game the shooting is let, or to let, by some one who claims to be lord of the manor, who lives at a distance or abroad. At all events he is not known personally to the people, and all they know about the overlordship is that, whereas in years gone by every villager had certain rights in the down—to cut furze and keep a cow, or pony, or donkey, or half a dozen sheep or goats—now they have none; but how and why and when these rights were lost nobody knows. Naturally there is no sympathy between the villagers and the keepers sent from a distance to protect the game, so that the shooting may be let to some other stranger. On the contrary, they religiously destroy every nest they can find, with the result that there are too few birds for anyone to take the shooting, and it remains year after year unlet.

This unsettled state of things is all to the advantage of the black and white bird with the ornamental tail, and he flourishes accordingly and builds his big, thorny nests in the roadside trees about the village.

The one big bird on these downs, as in so many other places in England, is the rook, and let us humbly thank the gods who own this green earth and all the creatures which inhabit it that they have in their goodness left us this one. For it is something to have a rook, although he is not a great bird compared with the great ones lost—bustard and kite and raven and goshawk, and many others. His abundance

on the cultivated downs is rather strange when one remembers the outcry made against him in some parts on account of his injurious habits; but here it appears the sentiment in his favour is just as strong in the farmer, or in a good many farmers, as in the great landlord. The biggest rookery I know on Salisbury Plain is at a farm-house where the farmer owns the land himself and cultivates about nine hundred acres. One would imagine that he would keep his rooks down in these days when a boy cannot be hired to scare the birds from the crops.

One day, near West Knoyle, I came upon a vast company of rooks busily engaged on a ploughed field where everything short of placing a bird-scarer on the ground had been done to keep the birds off. A score of rooks had been shot and suspended to long sticks planted about the field, and there were three formidable-looking men of straw and rags with hats on their heads and wooden guns under their arms. But the rooks were there all the same; I counted seven at one spot, prodding the earth close to the feet of one of the scare-crows. I went into the field to see what they were doing, and found that it was sown with vetches, just beginning to come up, and the birds were digging the seed up.

Three months later, near the same spot, on Mere Down, I found these birds feasting on the corn, when it had been long cut but could not be carried on account of the wet weather. It was a large field of fifty to sixty acres, and as I walked by it the birds came flying leisurely over my head to settle with loud cawings on the stooks. It was a magnificent sight—the great, blue-black bird-forms on the golden wheat, an animated group of three or four to half a dozen on every stook, while others walked about the ground to pick up the scattered grain, and others were flying over them, for just then the sun was shining on the field and beyond it the sky was blue. Never had I witnessed birds so manifestly rejoicing at their good fortune, with happy, loud caw-caw. Or rather haw-haw! what a harvest, what abundance! was there ever a more perfect August and September! Rain, rain, by night and in the morning; then sun and wind to dry our feathers and make us glad, but never enough to dry the corn to enable them to carry it and build it up in stacks where it would be so much harder to get at. Could anything be better!

But the commonest bird, the one which vastly outnumbers all the others I have named together, is the starling. It was Caleb Bawcombe's favourite bird, and I believe it is regarded with peculiar affection by all shepherds on the downs on account of its constant association with sheep in the pasture. The dog, the sheep, and the crowd of starlings —these are the lonely man's companions during his long days on the hills from April or May to November. And what a wise bird he is, and how well he knows his friends and his enemies! There was nothing more beautiful to see, Caleb would say, than the behaviour of a flock of starlings when a hawk was about. If it was a kestrel they took little or no notice of it, but if a sparrowhawk made its appearance, instantly the crowd of birds could be seen flying at furious speed towards the nearest flock of sheep, and down into the flock they would fall like a shower of stones and instantly disappear from sight. There they would remain on the ground, among the legs of the grazing sheep, until the hawk had gone on his way and passed out of sight.

The sparrowhawk's victims are mostly made among the young birds that flock together in summer and live apart from the adults during the summer months after the breeding season is over.

When I find a dead starling on the downs ranged over by sparrowhawks, it is almost always a young bird—a 'brown thrush' as it used to be called by the old naturalists. You may know that the slayer was a sparrowhawk by the appearance of the bird, its body untouched, but the flesh picked neatly from the neck and the head gone. That was swallowed whole, after the beak had been cut off. You will find the beak lying by the side of the body. In summer-time, when birds are most abundant, after the breeding season, the sparrowhawk is a fastidious feeder.

XI

STARLINGS AND SHEEP-BELLS

Starlings' singing—Native and borrowed sounds—Imitations of sheep-bells
—The shepherd on sheep-bells—The bells for pleasure, not use—A dog in
charge of the flock—Shepherd calling his sheep—Richard Warner of Bath
—Ploughmen singing to their oxen in Cornwall—A shepherd's loud singing

THE subject of starlings associating with sheep has served to remind me of something I have often thought when listening to their music. It happens that I am writing this chapter in a small village on Salisbury Plain, the time being mid-September, 1909, and that just outside my door there is a group of old elder-bushes laden just now with clusters of ripe berries on which the starlings come to feed, filling the room all day with that never-ending medley of sounds which is their song. They sing in this way not only when they sing—that is to say, when they make a serious business of it, standing motionless and a-shiver on the tiles, wings drooping and open beak pointing upwards, but also when they are feasting on fruit—singing and talking and swallowing elderberries between whiles to wet their whistles. If the weather is not too cold you will hear this music daily, wet or dry, all the year round. We may say that of all singing birds they are most vocal, yet have no set song. I doubt if they have more than half a dozen to a dozen sounds or notes which are the same in every individual and their very own. One of them is a clear, soft, musical whistle, slightly inflected; another a kissing sound, usually repeated two or three times or oftener, a somewhat percussive smack; still another, a sharp, prolonged hissing or sibilant but at the same time metallic note, compared by some one to the sound produced by milking a cow into a tin pail—a very good description. There are other lesser notes: a musical, thrush-like chirp, repeated slowly, and sometimes rapidly till it runs to a bubbling sound; also there is a horny sound, which is perhaps produced by striking upon the edges of the lower mandible with those of the upper. But it is quite unlike the loud, hard noise made by the stork; the poor stork being a dumb bird has made a sort of policeman's rattle of his huge beak. These sounds do not follow each other; they come from time to time, the intervals being filled up with others in such endless variety, each bird producing its own notes, that one can but suppose that they are imitations. We know, in fact, that the starling is our greatest mimic, and that he often succeeds in recognizable reproductions of single notes, of phrases, and occasionally of entire songs, as, for instance, that of the blackbird. But in listening to him we are conscious of his imitations; even when at his best he amuses rather than delights—he is not like the mocking-bird. His common starling pipe cannot produce sounds of pure and beautiful quality, like the blackbird's 'oboe-voice,' to quote Davidson's apt phrase; he emits this song in a strangely subdued tone, producing the effect of a blackbird heard singing at a considerable distance. And so with innumerable other notes, calls, and songs—they are often to their originals what a man's voice heard on a telephone is to his natural voice. He succeeds best, as a rule, in imitations of the coarser, metallic sounds, and as his medley abounds in a variety of little, measured, tinkling and clinking notes, as of tappings on a metal plate, it has struck me at times that these are probably borrowed from the sheep-bells of which the bird hears so much in his feeding-grounds. It is, however, not necessary to suppose that every starling gets these sounds directly from the bells; the birds undoubtedly mimic one another, as is the case with mocking-birds, and the young might easily acquire this part of their song language from the old birds without visiting the flocks in the pastures.

The sheep-bell, in its half-muffled strokes, as of a small hammer tapping on an iron or copper plate, is, one would imagine, a sound well within the starling's range, easily imitated, therefore specially attractive to him.

But—to pass to another subject—what does the shepherd himself think or feel about it; and why does he have bells on his sheep?

He thinks a great deal of his bells. He pipes not like the shepherd of fable or of the pastoral poets, nor plays upon any musical instrument, and seldom sings, or even whistles—that sorry substitute for song; he loves music nevertheless, and gets it in his sheep-bells; and he likes it in quantity. 'How many bells have you got on your sheep—it sounds as if you had a great many?' I asked of a shepherd the other day, feeding his flock near Old Sarum, and he replied,'Just forty, and I wish there were eighty.' Twenty-five or thirty is a more usual number, but only because of their cost, for the shepherd has very little money for bells or anything else. Another told me that he had 'only thirty,' but he intended getting more. The sound cheers him; it is not exactly monotonous, owing to the bells being of various sizes and also greatly varying in thickness, so that they produce different tones, from the sharp tinkle-tinkle of the smallest to the sonorous klonk-klonk of the big, copper bell. Then, too, they are differently agitated, some quietly when the sheep are grazing with heads down, others rapidly as the animal walks or trots on; and there are little bursts or peals when a sheep shakes its head, all together producing a kind of rude harmony —a music which, like that of bagpipes or of chiming church-bells, heard from a distance, is akin to natural music and accords with rural scenes.

As to use, there is little or none. A shepherd will sometimes say, when questioned on the subject, that the bells tell him just where the flock is or in which direction they are travelling; but he knows better. The one who is not afraid to confess the simple truth of the matter to a stranger will tell you that he does not need the bell to tell him where the sheep are or in which direction they are grazing. His eyes are good enough for that. The bells are for his solace or pleasure alone. It may be that the sheep like the tinkling too—it is his belief that they do like it. A shepherd said to me a few days ago: 'It is lonesome with the flock on the downs; more so in cold, wet weather, when you per- haps don't see a person all day—on some days not even at a distance, much less to speak to. The bells keep us from feeling it too much. We know what we have them for, and the more we have the better we like it. They are company to us.'

Even in fair weather he seldom has anyone to speak to. A visit from an idle man who will sit down and have a pipe and talk with him is a day to be long remembered and even to date events from. ' 'Twas the month—May, June, or October—when the stranger came out to the down and talked to I.'

One day, in September, when sauntering over Mere Down, one of the most extensive and loneliest-looking sheep-walks in South Wilts— a vast, elevated plain or table-land, a portion of which is known as White Sheet Hill—I passed three flocks of sheep, all with many bells, and noticed that each flock produced a distinctly different sound or effect, owing doubtless to a different number of big and little bells in each; and it struck me that any shepherd on a dark night, or if taken blindfolded over the downs, would be able to identify his own flock by the sound. At the last of the three flocks a curious thing occurred. There was no shepherd with it or anywhere in sight, but a dog was in charge; I found him lying apparently asleep in a hollow, by the side of a stick and an old sack. I called to him, but instead of jumping up and coming to me, as he would have done if his master had been there, he only raised his head, looked at me, then put his nose down on his paws again. I am on duty—in sole charge—and you must not speak to me, was what he said. After walking a little distance on, I spied the shepherd with a second dog at his heels, coming over the down straight to the flock, and I stayed to watch. When still over a hundred yards from the hollow the dog flew ahead, and the other jumping up ran to meet him, and they stood together, wagging their tails as if conversing. When the shepherd had got up to them he stood and began uttering a curious call, a somewhat musical cry in two notes, and instantly the sheep, now at a considerable distance, stopped feeding and turned, then all together began running towards him, and when within thirty yards stood still, massed together, and all gazing at him. He then uttered a different call, and turning walked away, the dogs keeping with him and the sheep closely following. It was late in the day, and he was going to fold them down at the foot of the slope in some fields half a mile away.

As the scene I had witnessed appeared unusual I related it to the very next shepherd I talked with.

'Oh, there was nothing in that,' he said. 'Of course the dog was behind the flock.'

I said, 'No, the peculiar thing was that both dogs were with their master, and the flock followed.'

'Well, my sheep would do the same,' he returned. 'That is, they'll do it if they know there's something good for them—something they like in the fold. They are very knowing.' And other shepherds to whom I related the incident said pretty much the same, but they apparently did not quite like to hear that any shepherd could control his sheep with his voice alone; their way of receiving the story confirmed me in the belief that I had witnessed something unusual.

Before concluding this short chapter I will leave the subject of the Wiltshire shepherd and his sheep to quote a remarkable passage about men singing to their cattle in Cornwall, from a work on that county by Richard Warner of Bath, once a well-known and prolific writer of topographical and other books. They are little known now, I fancy, but he was great in his day, which lasted from about the middle of the eighteenth to about the middle of the nineteenth century—at all events, he died in 1857, aged 94. But he was not great at first, and finding when nearing middle age that he was not prospering, he took to the Church and had several livings, some of them running concurrently, as was the fashion in those dark days. His topographical work included Walks in Wales, in Somerset, in Devon, Walks in many places, usually taken in a stage-coach or on horseback, containing nothing worth remembering except perhaps the one passage I have mentioned, which is as follows:—

'We had scarcely entered Cornwall before our attention was agreeably arrested by a practice connected with the agriculture of the people, which to us was entirely novel. The farmers judiciously employ the fine oxen of the country in ploughing, and other processes of husbandry, to which the strength of this useful animal can be employed"—the Rev. Richard Warner is tedious, but let us be patient and see what follows—'to which the strength of this useful animal can be employed; and while the hinds are thus driving their patient slaves along the furrows, they continually cheer them with conversation, denoting approbation and pleasure. This encouragement is conveyed to them in a sort of chaunt, of very agrreeable modulation, which, floating through the air from different distances, produces a striking effect both on the ear and imagination. The notes are few and simple, and when delivered by a clear, melodious voice, have something expressive of that tenderness and affection which man naturally entertains for the companions of his labours, in a *pastoral state* of society, when, feeling more forcibly his dependence upon domesticated animals for support, he gladly reciprocates with them kindness and protection for comfort and subsistence. This wild melody was to me, I confess, peculiarly affecting. It seemed to draw more closely the link of friendship between man and the humbler tribes of *fellow mortals*. It solaced my heart with the appearance of humanity, in a world of violence and in times of universal hostile rage; and it gladdened my fancy with the contemplation of those days of heavenly harmony, promised in the predictions of eternal truth, when man, freed at length from prejudice and passion, shall seek his happiness in cultivating the mild, the benevolent, and the merciful sensibilities of his nature; and when the animal world, catching the virtues of its lord and master, shall soften into gentleness and love; when the wolf' . . .

And so on, clause after clause, with others to be added, until the whole sentence becomes as long as a fishing-rod. But apart from the fiddlededee, is the thing he states believable? It is a charming picture, and one would like to know more about that 'chaunt,' that 'wild melody.' The passage aroused my curiosity when in Cornwall, as it had appeared to me that in no part of England are the domestic animals so little considered by their masters. The R.S.P.C.A. is practically unknown there, and when watching the doings of shepherds or drovers with their sheep the question has occurred to me, What would my Wiltshire shepherd friends say of such a scene if they had witnessed it? There is nothing in print which I can find to confirm Warner's observations, and if you inquire of very old men who have been all their lives on the soil they will tell you that there has never been such a custom in their time, nor have they ever heard of it as existing formerly. Warner's Tour through Cornwall is dated 1808.

I take it that he described a scene he actually witnessed, and that he jumped to the conclusion that it was a common custom for the ploughman to sing to his oxen. It is not unusual to find a man anywhere singing to his oxen, or horses, or sheep, if he has a voice and is fond of exercising it. I remember that in a former book—'Nature in Downland'—I described the sweet singing of a cow-boy when tending his cows on a heath near Trotton, in West Sussex; and here in Wiltshire it amused me to listen, at a vast distance, to the robust singing of a shepherd while following his flock on the great lonely downs above Chitterne. He was a sort of Tamagno of the downs, with a tremendous voice audible a mile away.

XII

THE SHEPHERD AND THE BIBLE

Dan'l Burdon, the treasure-seeker—The shepherd's feeling for the Bible—
Effect of the pastoral life—The shepherd's story of Isaac's boyhood—
The village on the Wylye

ONE of the shepherd's early memories was of Dan'l Burdon, a labourer
on the farm where Isaac Bawcombe was head-shepherd. He retained a
vivid recollection of this person, who had a profound gravity and was
the most silent man in the parish. He was always thinking about
hidden treasure, and all his spare time was spent in seeking for it. On a
Sunday morning, or in the evening after working hours, he would take
a spade or pick and go away over the hills on his endless search after
'something he could not find.' He opened some of the largest barrows,
making trenches six to ten feet deep through them, but found nothing
to reward him. One day he took Caleb with him, and they went to a
part of the down where there were certain depressions in the turf of
a circular form and six to seven feet in circumference. Burdon had
observed these basin-like depressions and had thought it possible they
marked the place where things of value had been buried in long-past
ages. To begin he cut the turf all round and carefully removed it, then
dug and found a thick layer of flints. These removed, he came upon
a deposit of ashes and charred wood. And that was all. Burdon without a word set to work to put it all back in its place again—ashes and
wood, and earth and flints—and having trod it firmly down he carefully replaced the turf, then leaning on his spade gazed silently at the
spot for a space of several minutes. At last he spoke. 'Maybe, Caleb,
you've heared tell about what the Bible says of burnt sacrifice. Well
now, I be of opinion that it were here. They people the Bible says
about, they come up here to sacrifice on White Bustard Down, and
these be the places where they made their fires.'

Then he shouldered his spade and started home, the boy following.
Caleb's comment was: 'I didn't say nothing to un because I were only
a leetel boy and he were a old man; but I knowed better than that all
the time, because them people in the Bible they was never in England

86

at all, so how could they sacrifice on White Bustard Down in Wiltsheer?'

It was no idle boast on his part. Caleb and his brothers had been taught their letters when small, and the Bible was their one book, which they read not only in the evenings at home but out on the downs during the day when they were with the flock. His extreme familiarity with the whole Scripture narrative was a marvel to me; it was also strange, considering how intelligent a man he was, that his lifelong reading of that one book had made no change in his rude 'Wiltsheer' speech.

Apart from the feeling which old, religious country people, who know nothing about the Higher Criticism, have for the Bible, taken literally as the Word of God, there is that in the old Scriptures which appeals in a special way to the solitary man who feeds his flock on the downs. I remember well in the days of my boyhood and youth, when living in a purely pastoral country among a semi-civilized and very simple people, how understandable and eloquent many of the ancient stories were to me. The life, the outlook, the rude customs, and the vivid faith in the Unseen, were much the same in that different race in a far-distant age, in a remote region of the earth, and in the people I mixed with in my own home. That country has been changed now; it has been improved and civilized and brought up to the European standard; I remember it when it was as it had existed for upwards of two centuries before it had caught the contagion. The people I knew were the descendants of the Spanish colonists of the seventeenth century, who had taken kindly to the life of the plains, and had easily shed the traditions and ways of thought of Europe and of towns. Their philosophy of life, their ideals, their morality, were the result of the conditions they existed in, and wholly unlike ours; and the conditions were like those of the ancient people of which the Bible tells us. Their very phraseology was strongly reminiscent of that of the sacred writings, and their character in the best specimens was like that of the men of the far past who lived nearer to God, as we say, and certainly nearer to nature than it is possible for us in this artificial state. Among these sometimes grand old men who were large landowners, rich in flocks and herds, these fine old, dignified 'natives,' the substantial and leading men of the district who could not spell their own names, there were those who reminded you of Abraham and Isaac and Jacob and Esau and Joseph and his brethren, and even of David the passionate psalmist, with perhaps a guitar for a harp.

No doubt the Scripture lessons read in the thousand churches on every Sunday of the year are practically meaningless to the hearers. These old men, with their sheep and goats and wives, and their talk about God, are altogether out of our ways of thought, in fact as far from us—as incredible or unimaginable, we may say—as the neolithic men or the inhabitants of another planet. They are of the order of mythical heroes and the giants of antiquity. To read about them is an ancient custom, but we do not listen.

Even to myself the memories of my young days came to be regarded as very little more than mere imaginations, and I almost ceased to believe in them until, after years of mixing with modern men, mostly in towns, I fell in with the downland shepherds, and discovered that even here, in densely populated and ultra-civilized England, something of the ancient spirit had survived. In Caleb, and a dozen old men more or less like him, I seemed to find myself among the people of the past, and sometimes they were so much like some of the remembered, old, sober, and slow-minded herders of the plains that I could not help saying to myself, Why, how this man reminds me of Tio Isidoro, or of Don Pascual of the 'Three Poplar Trees,' or of Marcos who would always have three black sheep in a flock. And just as they reminded me of these men I had actually known, so did they bring back the older men of the Bible history—Abraham and Jacob and the rest.

The point here is that these old Bible stories have a reality and significance for the shepherd of the down country which they have lost for modern minds; that they recognize their own spiritual lineaments in these antique portraits, and that all these strange events might have happened a few years ago and not far away.

One day I said to Caleb Bawcombe that his knowledge of the Bible, especially of the old part, was greater than that of the other shepherds I knew on the downs, and I would like to hear why it was so. This led to the telling of a fresh story about his father's boyhood, which he had heard in later years from his mother. Isaac was an only child and not the son of a shepherd; his father was a rather worthless if not a wholly bad man; he was idle and dissolute, and being remarkably dexterous with his fists he was persuaded by certain sporting persons to make a business of fighting—quite a common thing in those days. He wanted nothing better and spent the greater part of the time in wandering about the country; the money he made was spent away from home, mostly in drink, while his wife was left to keep herself and child in the best way she could at home or in the fields. By and by a poor stranger came to the village in search of work and was engaged for very little pay by a small farmer, for the stranger confessed that he was without experience of farm work of any description. The cheapest lodging he could find was in the poor woman's cottage, and then Isaac's mother, who pitied him because he was so poor and a stranger alone in the world, a very silent, melancholy man, formed the opinion that he had belonged to another rank in life. His speech and hands and personal habits betrayed it. Undoubtedly he was a gentleman; and then from something in his manner, his voice, and his words whenever he addressed her, and his attention to religion, she further concluded that he had been in the Church; that, owing to some trouble or disaster, he had abandoned his place in the world to live away from all who had known him, as a labourer.

One day he spoke to her about Isaac; he said he had been observing him and thought it a great pity that such a fine, intelligent boy should be allowed to grow up without learning his letters. She agreed that it was, but what could she do? The village school was kept by an old woman, and though she taught the children very little it had to be paid for, and she could not afford it. He then offered to teach Isaac himself and she gladly consented, and from that day he taught Isaac for a couple of hours every evening until the boy was able to read very well, after which they read the Bible through together, the poor man explaining everything, especially the historical parts, so clearly and beautifully, with such an intimate knowledge of the countries and peoples and customs of the remote East, that it was all more interesting than a fairy-tale. Finally he gave his copy of the Bible to Isaac, and told him to carry it in his pocket every day when he went out on the downs, and when he sat down to take it out and read in it. For by this time Isaac, who was now ten years old, had been engaged as a shepherd-boy to his great happiness, for to be a shepherd was his ambition.

Then one day the stranger rolled up his few belongings in a bundle and put them on a stick which he placed on his shoulder, said good-bye, and went away, never to return, taking his sad secret with him.

Isaac followed the stranger's counsel, and when he had sons of his own made them do as he had done from early boyhood. Caleb had never gone with his flock on the down without the book, and had never passed a day without reading a portion.

The incidents and observations gathered in many talks with the old shepherd, which I have woven into the foregoing chapters, relate mainly to the earlier part of his life, up to the time when, a married man and father of three small children, he migrated to Warminster. There he was in, to him, a strange land, far away from friends and home and the old, familiar surroundings, amid new scenes and new people. But the few years he spent at that place had furnished him with many interesting memories, some of which will be narrated in the following chapters.

I have told in the account of Winterbourne Bishop how I first went to that village just to see his native place, and later I visited Doveton for no other reason than that he had lived there, to find it one of the most charming of the numerous pretty villages in the vale. I looked for the cottage in which he had lived and thought it as perfect a home as a quiet, contemplative man who loved nature could have had: a small thatched cottage, very old looking, perhaps inconvenient to live in, but situated in the prettiest spot, away from other houses, near and within sight of the old church with old elms and beech-trees growing close to it, and the land about it green meadow. The clear river, fringed with a luxuriant growth of sedges, flag, and reeds, was less than a stone's-throw away.

So much did I like the vale of the Wylye when I grew to know it well that I wish to describe it fully in the chapter that follows.

XIII

VALE OF THE WYLYE

Warminster—Vale of the Wylye—Counting the villages—A lost church—
Character of the villages—Tytherington church—Story of the dog—Lord
Lovell—Monuments in churches—Manor-houses—Knook—The cottages—
Yellow stonecrop—Cottage gardens—Marigolds—Golden-rod—Wild flowers
of the water-side—Seeking for the characteristic expression

THE PRETTILY-NAMED Wylye is a little river not above twenty miles
in length from its rise to Salisbury, where after mixing with the Nadder
at Wilton, it joins the Avon. At or near its source stands Warminster, a
small, unimportant town with a nobler-sounding name than any other
in Wiltshire. Trowbridge, Devizes, Marlborough, Salisbury, do not stir
the mind in the same degree; and as for Chippenham, Melksham, Mere,
Calne, and Corsham, these all are of no more account than so many
villages in comparison. Yet Warminster has no associations—no place
in our mental geography; at all events one remembers nothing about
it. Its name, which after all may mean nothing more than the monastery on the Were—one of the three streamlets which flow into the
Wylye at its source—is its only glory. It is not surprising that Caleb
Bawcombe invariably speaks of his migration to, and of the time he
passed at Warminster, when, as a fact, he was not there at all, but at
Doveton, a little village on the Wylye a few miles below the town
with the great name.

It is a green valley—the greenness strikes one sharply on account of
the pale colour of the smooth, high downs on either side—half a mile
to a mile in width, its crystal current showing like a bright serpent
for a brief space in the green, flat meadows, then vanishing again
among the trees. So many are the great shade trees, beeches and ashes
and elms, that from some points the valley has the appearance of a
continuous wood—a contiguity of shade. And the wood hides the villages, at some points so effectually that looking down from the hills
you may not catch a glimpse of one and imagine it to be a valley
where no man dwells. As a rule you do see something of human occu-
pancy—the red or yellow roofs of two or three cottages, a half hidden
grey church tower, or column of blue smoke, but to see the villages

you must go down and look closely, and even so you will find it difficult to count them all. I have tried, going up and down the valley several times, walking or cycling, and have never succeeded in getting the
same number on two occasions. There are certainly more than twenty,
without counting the hamlets, and the right number is probably something between twenty-five and thirty, but I do not want to find out by
studying books and maps. I prefer to let the matter remain unsettled so
as to have the pleasure of counting or trying to count them again at
some future time. But I doubt that I shall ever succeed. On one occasion
I caught sight of a quaint, pretty little church standing by itself in the
middle of a green meadow, where it looked very solitary with no
houses in sight and not even a cow grazing near it. The river was between me and the church, so I went up-stream, a mile and a half, to
cross by the bridge, then doubled back to look for the church, and
couldn't find it! Yet it was no illusory church; I have seen it again on
two occasions, but again from the other side of the river, and I must
certainly go back some day in search of that lost church, where there
may be effigies, brasses, sad, eloquent inscriptions, and other memorials
of ancient tragedies and great families now extinct in the land.

This is perhaps one of the principal charms of the Wylye—the
sense of beautiful human things hidden from sight among the masses
of foliage. Yet another lies in the character of the villages. Twenty-
five or twenty-eight of them in a space of twenty miles; yet the impression left on the mind is that these small centres of population are
really few and far between. For not only are they small, but of the
old, quiet, now almost obsolete type of village, so unobtrusive as to
affect the mind soothingly, like the sight of trees and flowery banks
and grazing cattle. The churches, too, as is fit, are mostly small and
ancient and beautiful, half-hidden in their tree-shaded churchyards,
rich in associations which go back to a time when history fades into
myth and legend. Not all, however, are of this description; a few are
naked, dreary little buildings, and of these I will mention one which,
albeit ancient, has no monuments and no burial-ground. This is the
church of Tytherington, a small, rustic village, which has for neighbours Codford St. Peter on one side and Sutton Veny and Norton
Bavant on the other. To get to this church, where there was nothing
but naked walls to look at, I had to procure the key from the clerk, a
nearly blind old man of eighty. He told me that he was shoemaker but
could no longer see to make or mend shoes; that as a boy he was a

weak, sickly creature, and his father, a farm bailiff, made him learn shoemaking because he was unfit to work out of doors. 'I remember this church,' he said, 'when there was only one service each quarter,' but, strange to say, he forgot to tell me the story of the dog! 'What, didn't he tell you about the dog?' exclaimed everybody. There was really nothing else to tell.

It happened about a hundred years ago that once, after the quarterly service had been held, a dog was missed, a small terrier owned by the young wife of a farmer of Tytherington named Case. She was fond of her dog, and lamented its loss for a little while, then forgot all about it. But after three months, when the key was once more put into the rusty lock and the door thrown open, there was the dog, a living 'skelington' it was said, dazed by the light of day, but still able to walk! It was supposed that he had kept himself alive by 'licking the moisture from the walls.' The walls, they said, were dripping with wet and covered with a thick growth of mould. I went back to interrogate the ancient clerk, and he said that the dog died shortly after its deliverance; Mrs. Case herself told him all about it. She was an old woman then, but was always willing to relate the sad story of her pet.

That picture of the starving dog coming out, a living skeleton, from the wet, mouldy church, reminds us sharply of the changed times we live in and of the days when the Church was still sleeping very peacefully, not yet turning uneasily in its bed before opening its eyes; and when a comfortable rector of Codford thought it quite enough that the people of Tytherington, a mile away, should have one service every three months.

As a fact, the Tytherington dog interested me as much as the story of the last Lord Lovell's self-incarceration in his own house in the neighbouring little village of Upton Lovell. He took refuge there from his enemies who were seeking his life, and concealed himself so effectually that he was never seen again. Centuries later, when excavations were made on the site of the ruined mansion, a secret chamber was discovered, containing a human skeleton seated in a chair at a table, on which were books and papers crumbling into dust.

A volume might be filled with such strange and romantic happenings in the little villages of the Wylye, and for the natural man they have a lasting fascination; but they invariably relate to great people of their day—warriors and statesmen and landowners of old and noble lineage, the smallest and meanest you will find being clothiers, or merchants, who amassed large fortunes and built mansions for themselves and almshouses for the aged poor, and, when dead, had memorials placed to them in the churches. But of the humble cottagers, the true people of the vale who were rooted in the soil, and flourished and died like trees in the same place—of these no memory exists. We only know that they lived and laboured; that when they died, three or four a year, three or four hundred in a century, they were buried in the little shady churchyard, each with a green mound over him to mark the spot. But in time these 'mouldering heaps' subsided, the bodies turned to dust, and another and yet other generations were laid in the same place among the forgotten dead, to be themselves in turn forgotten. Yet I would rather know the histories of these humble, unremembered lives than of the great ones of the vale who have left us a memory.

It may be for this reason that I was little interested in the manor-houses of the vale. They are plentiful enough, some gone to decay or put to various uses; others still the homes of luxury, beauty, culture: stately rooms, rich fabrics; pictures, books, and manuscripts, gold and silver ware, china and glass, expensive curios, suits of armour, ivory and antlers, tiger-skins, stuffed goshawks and peacocks' feathers. Houses, in some cases built centuries ago, standing half-hidden in beautiful wooded grounds, isolated from the village; and even as they thus stand apart, sacred from intrusion, so the life that is in them does not mix with or form part of the true native life. They are to the cottagers of to-day what the Roman villas were to the native population of some eighteen centuries ago. This will seem incredible to some: to me, an untrammelled person, familiar in both hall and cottage, the distance between them appears immense.

A reader well acquainted with the valley will probably laugh to be told that the manor-house which most interested me was that of Knook, a poor little village between Heytesbury and Upton Lovell. Its ancient and towerless little church with rough, grey walls is, if possible, even more desolate-looking than that of Tytherington. In my hunt for the key to open it I disturbed a quaint old man, another octogenarian, picturesque in a vast white beard, who told me he was a thatcher, or had been one before the evil days came when he could work no more and was compelled to seek parish relief. 'You must go to the manor-house for the key,' he told me. A strange place in which to look for the key, and it was stranger still to see the house, close to the

church, and so like it that but for the small cross on the roof of the latter one could not have known which was the sacred building. First a
monks' house, it fell at the Reformation to some greedy gentleman
who made it his dwelling, and doubtless in later times it was used as a
farm-house. Now a house most desolate, dirty, and neglected, with
cracks in the walls which threaten ruin, standing in a wilderness of
weeds tenanted by a poor working-man whose wages are twelve shillings a week, and his wife and eight small children. The rent is eighteen-
pence a week—probably the lowest-rented manor-house in England,
though it is not very rare to find such places tenanted by labourers.

But let us look at the true cottages. There are, I imagine, few places
in England where the humble homes of the people have so great a
charm. Undoubtedly they are darker inside, and not so convenient to
live in as the modern box-shaped, red-brick, slate-roofed cottages,
which have spread a wave of ugliness over the country; but they do
not offend—they please the eye. They are smaller than the modern-built
habitations; they are weathered and coloured by sun and wind and rain
and many lowly vegetable forms to a harmony with nature. They
appear related to the trees amid which they stand, to the river and
meadows, to the sloping downs at the side, and to the sky and clouds
over all. And, most delightful feature, they stand among, and are wrapped in, flowers as in a garment—rose and vine and creeper and clematis.
They are mostly thatched, but some have tiled roofs, their deep, dark
red clouded and stained with lichen and moss; and these roofs, too,
have their flowers in summer. They are grown over with yellow stonecrop, that bright cheerful flower that smiles down at you from the
lowly roof above the door, with such an inviting expression, so delighted to see you no matter how poor and worthless a person you
may be or what mischief you may have been at, that you begin to
understand the significance of a strange vernacular name of this plant
—Welcome-home-husband-though-never-so-drunk.

But its garden flowers, clustering and nestling round it, amid which
its feet are set—they are to me the best of all flowers. These are the
flowers we know and remember for ever. The old, homely, cottage-
garden blooms, so old that they have entered the soul. The big house
garden, or gardener's garden, with everything growing in it I hate, but
these I love—fragrant gilly-flower and pink and clove-smelling carna-
tion; wallflower, abundant periwinkle, sweet-william, larkspur, love-
in-a-mist, and love-lies-bleeding, old-woman's-nightcap, and kiss-me-

95

John-at-the-garden-gate, sometimes called pansy. And best of all and in greatest profusion, that flower of flowers, the marigold.

How the townsman, town born and bred, regards this flower, I do not know. He is, in spite of all the time I have spent in his company, a comparative stranger to me—the one living creature on the earth who does not greatly interest me. Some over-populated planet in our system discovered a way to relieve itself by discharging its superfluous millions on our globe—a pale people with hurrying feet and eager, restless minds, who live apart in monstrous, crowded camps, like wood ants that go not out to forage for themselves—six millions of them crowded together in one camp alone! I have lived in these colonies, years and years, never losing the sense of captivity, of exile, ever concious of my burden, taking no interest in the doings of that innumerable multitude, its manifold interests, its ideals and philosophy, its arts and pleasures. What, then, does it matter how they regard this common orange-coloured flower with a strong smell? For me it has an atmosphere, a sense or suggestion of something immeasurably remote and very beautiful—an event, a place, a dream perhaps, which has left no distinct image, but only this feeling unlike all others, imperishable, and not to be described except by the one word Marigold.

But when my sight wanders away from the flower to others blooming with it—to all those which I have named and to the taller ones, so tall that they reach half-way up, and some even quite up, to the eaves of the lowly houses they stand against—hollyhocks and peonies and crystalline white lilies with powdery gold inside, and the common sunflower—I begin to perceive that they all possess something of that same magical quality.

These taller blooms remind me that the evening primrose, long naturalised in our hearts, is another common and very delightful cottage-garden flower; also that here, on the Wylye, there is yet another stranger from the same western world which is fast winning our affections. This is the golden-rod, grandly beautiful in its great, yellow, plume-like tufts. But it is not quite right to call the tufts yellow: they are green, thickly powdered with the minute golden florets. There is no flower in England like it, and it is a happiness to know that it promises to establish itself with us as a wild flower.

Where the village lies low in the valley and the cottage is near the water, there are wild blooms, too, which almost rival those of the garden in beauty—water agrimony and comfrey with ivory-white and dim purple blossoms, purple and yellow loosestrife and gem-like, water forget-me-not; all these mixed with reeds and sedges and water-grasses, forming a fringe or border to the potato or cabbage patch, dividing it from the stream.

But now I have exhausted the subject of the flowers, and enumerated and dwelt upon the various other components of the scene, it comes to me that I have not yet said the right thing and given the Wylye its characteristic expression. In considering the flowers we lose sight of the downs, and so in occupying ourselves with the details we miss the general effect. Let me then, once more, before concluding this chapter, try to capture the secret of this little river.

There are other chalk streams in Wiltshire and Hampshire and Dorset—swift crystal currents that play all summer long with the floating poa grass fast held in their pebbly beds, flowing through smooth downs, with small ancient churches in their green villages, and pretty thatched cottages smothered in flowers—which yet do not produce the same effect as the Wylye. Not Avon for all its beauty, nor Itchen, nor Test. Wherein, then, does the 'Wylye bourne' differ from these others, and what is its special attraction? It was only when I set myself to think about it, to analyse the feeling in my own mind, that I discovered the secret—that is, in my own case, for of its effect on others I cannot say anything. What I discovered was that the various elements of interest, all of which may be found in other chalk-stream valleys, are here concentrated, or comprised in a limited space, and seen together produce a combined effect on the mind. It is the narrowness of the valley and the nearness of the high downs standing over it on either side, with, at some points, the memorials of antiquity carved on their smooth surfaces, the barrows and lynchetts or terraces, and the vast green earth-works crowning their summit. Up here on the turf, even with the lark singing his shrill music in the blue heavens, you are with the prehistoric dead, yourself for the time one of that innumerable, unsubstantial multitude, invisible in the sun, so that the sheep travelling as they graze, and the shepherd following them, pass through their ranks without suspecting their presence. And from that elevation you look down upon the life of to-day—the visible life, so brief in the individual, which, like the swift silver stream beneath, yet flows on continuously from age to age and for ever. And even as you look down you hear, at that distance, the bell of the little hidden church tower telling the hour of noon, and quickly following, a shout of freedom and joy from many shrill voices of children just released from school. Woke to life by those sounds, and drawn down by them, you may sit to rest or sun yourself on the stone table of a tomb overgrown on its sides with moss, the two-century-old inscription wellnigh obliterated, in the little grass-grown, flowery churchyard which serves as village green and playground in that small centre of life, where the living and the dead exist in a neighbourly way together. For it is not here as in towns, where the dead are away and out of mind and the past cut off. And if after basking too long in the sun in that tree-sheltered spot you go into the little church to cool yourself, you will probably find in a dim corner not far from the altar a stone effigy of one of an older time; a knight in armour, perhaps a crusader with legs crossed, lying on his back, dimly seen in the dim light, with perhaps a coloured sunbeam on his upturned face. For this little church where the villagers worship is very old; Norman on Saxon foundations; and before they were ever laid there may have been a temple to some ancient god at that spot, or a Roman villa perhaps. For older than Saxon foundations are found in the vale, and mosaic floors, still beautiful after lying buried so long.

All this—the far-removed events and periods in time—are not in the conscious mind when we are in the vale or when we are looking down on it from above: the mind is occupied with nothing but visible nature. Thus, when I am sitting on the tomb, listening to the various sounds of life about me, attentive to the flowers and bees and butterflies, to man or woman or child taking a short cut through the churchyard, exchanging a few words with them; or when I am by the water close by, watching a little company of graylings, their delicately-shaded, silver-grey scales distinctly seen as they lie in the crystal current watching for flies; or when I listen to the perpetual musical talk and song combined of a family of greenfinches in the alders or willows, my mind is engaged with these things. But if one is familiar with the vale; if one has looked with interest and been deeply impressed with the signs and memorials of past life and of antiquity everywhere present and forming part of the scene, something of it and of all that it represents remains in the subconscious mind to give a significance and feeling to the scene, which affects us here more than in most places; and that, I take it, is the special charm of this little valley.

XIV

A SHEEP-DOG'S LIFE

Watch—His visits to a dew-pond—David and his dog Monk—Watch goes to David's assistance—Caleb's new master objects to his dog—Watch and the corn-crake—Watch plays with rabbits and guinea-pigs—Old Nance the rook-scarer—The lost pair of spectacles—Watch in decline—Grey hairs in animals—A grey mole—Last days of Watch—A shepherd on old sheep-dogs

PERHAPS the most interesting of the many sheep-dog histories the shepherd related was that of Watch, a dog he had at Winterbourne Bishop for three years before he migrated to Warminster. Watch, he said, was more 'like a Christian,' otherwise a reasonable being, than any other dog he had owned. He was exceedingly active, and in hot weather suffered more from heat than most dogs. Now the only accessible water when they were out on the down was in the mist-pond about a quarter of a mile from his 'liberty,' as he called that portion of the down on which he was entitled to pasture his sheep. When Watch could stand his sufferings no longer, he would run to his master, and sitting at his feet look up at his face and emit a low, pleading whine.

'What be you wanting, Watch—a drink or a swim?' the shepherd would say, and Watch, cocking up his ears, would repeat the whine.

'Very well, go to the pond,' Bawcombe would say, and off Watch would rush, never pausing until he got to the water, and dashing in he would swim round and round, lapping the water as he bathed.

At the side of the pond there was a large, round sarsen-stone, and invariably on coming out of his bath Watch would jump upon it, and with his four feet drawn up close together would turn round and round, surveying the country from that elevation; then jumping down he would return in all haste to his duties.

Another anecdote, which relates to the Winterbourne Bishop period, is a somewhat painful one, and is partly about Monk, the sheep-dog already described as a hunter of foxes, and his tragic end. Caleb had worked him for a time, but when he came into possession of Watch he gave Monk to his younger brother David, who was under-shepherd on the same farm.

One morning Caleb was with the ewes in a field, when David, who was in charge of the lambs two or three fields away, came to him looking very strange—very much put out.

'What are you here for—what's wrong with 'ee?' demanded Caleb.

'Nothing's wrong,' returned the other.

'Where's Monk then?' asked Caleb.

'Dead,' said David.

'Dead! How's he dead?'

'I killed'n. He wouldn't mind me and made me mad, and I up with my stick and gave him one crack on the head and it killed'n.'

'You killed'n!' exclaimed Caleb. 'An' you come here an' tell I nothing's wrong! Is that a right way to speak of such a thing as that? What be you thinking of? And what be you going to do with the lambs?'

'I'm just going back to them—I'm going to do without a dog. I'm going to put them in the rape and they'll be all right.'

'What! put them in the rape and no dog to help 'ee?' cried the other. 'You are not doing things right, but master mustn't pay for it. Take Watch to help 'ee—I must do without'n this morning.'

'No, I'll not take'n,' he said, for he was angry because he had done an evil thing and he would have no one, man or dog, to help him. 'I'll do better without a dog,' he said, and marched off.

Caleb cried after him: 'If you won't have the dog don't let the lambs suffer but do as I tell 'ee. Don't you let 'em bide in the rape more'n ten minutes; then chase them out, and let 'em stand twenty minutes to half an hour; then let them in another ten minutes and out again for twenty minutes, then let them go back and feed in it quietly, for the danger'll be over. If you don't do as I tell 'ee you'll have many blown.'

David listened, then without a word went his way. But Caleb was still much troubled in his mind. How would he get that flock of hungry lambs out of the rape without a dog? And presently he determined to send Watch, or try to send him, to save the situation. David had been gone half an hour when he called the dog, and pointing in the direction he had taken he cried, 'Dave wants 'ee—go to Dave.'

Watch looked at him and listened, then bounded away, and after running full speed about fifty yards stopped to look back to make sure he was doing the right thing. 'Go to Dave,' shouted Caleb once more; and away went Watch again, and arriving at a very high gate at the end of the field dashed at and tried two or three times to get over it, first by jumping, then by climbing, and falling back each time. But by and by he managed to force his way through the thick hedge and was gone from sight.

When David came back that evening he was in a different mood, and said that Watch had saved him from a great misfortune: he could never have got the lambs out by himself, as they were mad for the rape. For some days after this Watch served two masters. Caleb would take him to his ewes, and after a while would say, 'Go—Dave wants 'ee,' and away Watch would go to the other shepherd and flock.

When Bawcombe had taken up his new place at Doveton, his master, Mr. Ellerby, watched him for a while with sharp eyes, but he was soon convinced the he had not made a mistake in engaging a head-shepherd twenty-five miles away without making the usual inquiries but merely on the strength of something heard casually in conversation about this man. But while more than satisfied with the man he remained suspicious of the dog. 'I'm afraid that dog of yours must hurt the sheep,' he would say, and even advised him to change him for one that worked in a quieter manner. Watch was too excitable, too impetuous—he could not go after the sheep in that violent way and grab them as he did without injuring them with his teeth.

'He did never bite a sheep in his life,' Bawcombe assured him, and eventually he was able to convince his master that Watch could make a great show of biting the sheep without doing them the least hurt—that it was actually against his nature to bite or injure anything.

One day in the late summer, when the corn had been cut but not carried, Bawcombe was with his flock on the edge of a newly reaped cornfield in a continuous, heavy rain, when he spied his master coming to him. He was in a very light summer suit and straw hat, and had no umbrella or other protection from the pouring rain. 'What be wrong with master to-day?' said Bawcombe. 'He's tarrably upset to be out like this in such a rain in a straw hat and no coat.'

Mr. Ellerby had by that time got into the habit when troubled in his mind of going out to his shepherd to have a long talk with him. Not a talk about his trouble—that was some secret bitterness in his heart—but just about the sheep and other ordinary topics, and the talk, Caleb said, would seem to do him good. But this habit he had got into was observed by others, and the farm-men would say, 'Something's wrong to-day—the master's gone off to the head-shepherd.'

When he came to where Bawcombe was standing, in a poor shelter by the side of a fence, he at once started talking on indifferent subjects, standing there quite unconcerned, as if he didn't even know that it was raining, though his thin clothes were wet through, and the water coming through his straw hat was running in streaks down his face. By and by he became interested in the dog's movements, playing about in the rain among the stooks. 'What has he got in his mouth?' he asked presently.

'Come here, Watch,' the shepherd called, and when Watch came he bent down and took a corncrake from his mouth. He had found the bird hiding in one of the stooks and had captured without injuring it.

'Why, it's alive—the dog hasn't hurt it,' said the farmer, taking it in his hands to examine it.

'Watch never hurted any creature yet,' said Bawcombe. He caught things just for his own amusement, but never injured them—he always let them go again. He would hunt mice in the fields, and when he captured one he would play with it like a cat, tossing it from him, then dashing after and recapturing it. Finally, he would let it go. He played with rabbits in the same way, and if you took a rabbit from him and examined it you would find it quite uninjured.

The farmer said it was wonderful—he had never heard of a case like it before; and talking of Watch he suceeded in forgetting the trouble in his mind which had sent him out in the rain in his thin clothes and straw hat, and he went away in a cheerful mood.

Caleb probably forgot to mention during this conversation with his master that in most cases when Watch captured a rabbit he took it to his master and gave it into his hands, as much as to say, Here is a very big sort of field-mouse I have caught, rather difficult to manage—perhaps *you* can do something with it?

The shepherd had many other stories about this curious disposition of his dog. When he had been some months in his new place his brother David followed him to the Wylye, having obtained a place as shepherd on a farm adjoining Mr. Ellerby's. His cottage was a little out of the village and had some ground to it, with a nice lawn or green patch. David was fond of keeping animal pets—birds in cages, and rabbits and guinea-pigs in hutches, the last so tame that he would release them on the grass to see them play with one another. When Watch first saw these pets he was very much attracted, and wanted to get to them, and after a good deal of persuasion on the part of Caleb, David one day consented to take them out and put them on the grass in the dog's presence. They were a little alarmed at first, but in a surprisingly short time made the discovery that this particular dog was not their enemy but a playmate. He rolled on the grass among them, and chased them round and round, and sometimes caught and pretended to worry them, and they appeared to think it very good fun.

'Watch,' said Bawcombe, 'in the fifteen years I had'n, never killed and never hurt a creature, no, not even a leetel mouse, and when he caught anything 'twere only to play with it.'

Watch comes into a story of an old woman employed at the farm at this period. She had been in the Warminster workhouse for a short time, and had there heard that a daughter of a former mistress in another part of the county had long been married and was now the mistress of Doveton Farm, close by. Old Nance thereupon obtained her release and trudged to Doveton, and one very rough, cold day presented herself at the farm to beg for something to do which would enable her to keep herself. If there was nothing for her she must, she said, go back and end her days in the Warminster workhouse. Mrs. Ellerby remembered and pitied her, and going in to her husband begged him earnestly to find some place on the farm for the forlorn old creature. He did not see what could be done for her: they already had one old woman on their hands, who mended sacks and did a few other trifling things, but for another old woman there would be nothing to do. Then he went in and had a good long look at her, revolving the matter in his mind, anxious to please his wife, and finally, he asked her if she could scare the crows. He could think of nothing else. Of course she could scare crows—it was the very thing for her! Well, he said, she could go and look after the swedes; the rooks had just taken a liking to them, and even if she was not very active perhaps she would be able to keep them off.

Old Nance got up to go and begin her duties at once. Then the farmer, looking at her clothes, said he would give her something more to protect her from the weather on such a bleak day. He got her an old felt hat, a big old, frieze overcoat, and a pair of old leather leggings. When she had put on these somewhat cumbrous things, and had tied her hat firmly on with a strip of cloth, and fastened the coat at the waist with a cord, she was told to go to the head-shepherd and ask him to direct her to the field where the rooks were troublesome. Then when she was setting out the farmer called her back and gave her an ancient, rusty gun to scare the birds. 'It isn't loaded,' he said, with a grim smile. 'I don't allow powder and shot, but if you'll point it at them they'll fly fast enough.'

Thus arrayed and armed she set forth, and Caleb seeing her approach at a distance was amazed at her grotesque appearance, and even more amazed still when she explained who and what she was and asked him to direct her to the field of swedes.

Some hours later the farmer came to him and asked him casually if he had seen an old gallus-crow about.

'Well,' replied the shepherd, 'I seen an old woman in man's coat and things, with an old gun, and I did tell she where to bide.'

'I think it will be rather cold for the old body in that field,' said the farmer. 'I'd like you to get a couple of padded hurdles and put them up for a shelter for her.'

And in the shelter of the padded or thatched hurdles, by the hedge-side, old Nance spent her days keeping guard over the turnips, and afterwards something else was found for her to do, and in the meanwhile she lodged in Caleb's cottage and became like one of the family. She was fond of the children and of the dog, and Watch became so much attached to her that had it not been for his duties with the flock he would have attended her all day in the fields to help her with the crows.

Old Nance had two possessions she greatly prized—a book and a pair of spectacles, and it was her custom to spend the day sitting, spectacles on nose and book in hand, reading among the turnips. Her spectacles were so 'tarrable' good that they suited all old eyes, and when this was discovered they were in great request in the village, and every person who wanted to do a bit of fine sewing or anything requiring young vision in old eyes would borrow them for the purpose. One day the old woman returned full of trouble from the fields—she had lost her spectacles; she must, she thought, have lent them to some one in the village on the previous evening and then forgotten all about it. But no one had them, and the mysterious loss of the spectacles was discussed and lamented by everybody. A day or two later Caleb came through the turnips on his way home, the dog at his heels, and when he got to his cottage Watch came round and placed himself square before his master and deposited the lost spectacles at his feet. He had found them in the turnip-field over a mile from home, and though but a dog he remembered that he had seen them on people's noses and in their hands, and knew that they must therefore be valuable—not to himself, but to that larger and more important kind of dog that goes about on its hind legs.

There is always a sad chapter in the life-history of a dog; it is the last one, which tells of his decline; and it is ever saddest in the case of the sheep-dog, because he has lived closer to man and has served him every day of his life with all his powers, all his intelligence, in the one useful and necessary work he is fitted for or which we have found for him to do. The hunting and the pet, or parasite, dogs—the 'dogs for sport and pleasure'—though one in species with him are not like beings of the same order; they are like professional athletes and per- formers, and smart or fashionable people compared to those who do the work of the world—who feed us and clothe us. We are accustomed to speak of dogs generally as the servants and the friends of man; it is only of the sheep-dog that this can be said with absolute truth. Not only is he the faithful servant of the solitary man who shepherds his flock, but the dog's companionship is as much to him as that of a fel- low-being would be.

Before his long and strenuous life was finished, Watch, originally jet-black without a spot, became quite grey, the greyness being most marked on the head, which became at last almost white.

It is undoubtedly the case that some animals, like men, turn grey with age, and Watch when fifteen was relatively as old as a man at sixty-five or seventy. But grey hairs do not invariably come with age, even in our domestic animals, which are more subject to this change than those in a state of nature. But we are never so well able to judge of this in the case of wild animals, as in most cases their lives end pre- maturely.

The shepherd related a curious instance in a mole. He once noticed mole-heaps of a peculiar kind in a field of sainfoin, and it looked to him as if this mole worked in a way of his own, quite unlike the others. The hills he threw up were a good distance apart, and so large that you could fill a bushel measure with the mould from any one of them. He noticed that this mole went on burrowing every day in the same manner; every morning there were new chains or ranges of the huge mounds. The runs were very deep, as he found when setting a mole-trap—over two feet beneath the surface. He set his trap, filling the deep hole he had made with sods, and on opening it next day he found his mole and was astonished at its great size. He took no measurements, but it was bigger, he affirmed, than he could have believed it possible for a mole to be. And it was grey instead of black, the grey hairs being so abundant on the head as to make it almost white, as in the case of old Watch. He supposed that it was a very old mole, that it was a more powerful digger than most of its kind, and had perhaps escaped death so long on account of its strength and of its habit of feeding deeper in the earth than the others.

To return to Watch. His hearing and eyesight failed as he grew older until he was practically blind and too deaf to hear any word given in the ordinary way. But he continued strong as ever on his legs, and his mind was not decayed, nor was he in the least tired. On the contrary, he was always eager to work, and as his blindness and deafness had made him sharper in other ways he was still able to make himself useful with the sheep. Whenever the hurdles were shifted to a fresh place and the sheep had to be kept in a corner of the enclosure until the new place was ready for them, it was old Watch's duty to keep them from breaking away. He could not see nor hear, but in some mysterious way he knew when they tried to get out, even if it was but one. Possibly the slight vibration of the ground informed him of the movement and the direction as well. He would make a dash and drive the sheep back, then run up and down before the flock until all was quiet again. But at last it became painful to witness his efforts, especially when the sheep were very restless, and incessantly trying to break away; and Watch finding them so hard to restrain would grow angry and rush at them with such fury that he would come violently against the hurdles at one side, then getting up, howling with pain, he would dash to the other side, when he would strike the hurdles there and cry out with pain once more.

It could not be allowed to go on; yet Watch could not endure to be deprived of his work; if left at home he would spend the time whining and moaning, praying to be allowed to go to the flock, until at last his master with a very heavy heart was compelled to have him put to death.

This is indeed almost invariably the end of a sheep-dog; however zealous and faithful he may have been, and however much valued and loved, he must at last be put to death. I related the story of this dog to a shepherd in the very district where Watch had lived and served his master so well—one who had been head-shepherd for upwards of forty years at Imber Court, the principal farm at the small downland village of Imber. He told me that during all his shepherding years he had never owned a dog which had passed out of his hands to another; every dog had been acquired as a pup and trained by himself; and he had been very fond of his dogs, but had always been compelled to have them shot in the end. Not because he would have found them too great a burden when they had become too old and their senses decayed, but because it was painful to see them in their decline, perpetually craving to be at their old work with the sheep, incapable of doing it any longer, yet miserable if kept from it.

XV

CONCERNING CATS

ONE OF the shepherd's most interesting memories of his Doveton period was of a cat they possessed, which was greatly admired. He was a very large, handsome, finely marked tabby, with a thick coat, and always appeared very well nourished but never wanted to be fed. He was a nice-tempered, friendly animal, and whenever he came in he appeared pleased at seeing the inmates of the house, and would go from one to another, rubbing his sides against their legs and purring aloud with satisfaction. Then they would give him food, and he would take a morsel or two or lap up as much milk as would fill a teaspoon and leave the rest. He was not hungry, and it always appeared, they said, as if he smelt at or tasted the food they put down for him just to please them. Everybody in the village admired their cat for his great size, his beauty, and gentle, friendly disposition, but how he fed himself was a mystery to all, since no one had ever detected him trying to catch anything out of doors. In that part of the Wylye valley there were no woods for him to hunt in; they also noticed that when, out of doors, the small birds, anxious and angered at his presence, would flit, uttering their cries, close to him, he paid no attention. The only thing they discovered about his outdoor life by watching him was that he had the habit of going to the railway track, then recently constructed, from Westbury to Salisbury, which ran near their cottage, and would there seat himself on one of the rails and remain for a long time gazing fixedly before him as if he found it a pleasure to keep his eyes on the long, glittering metal line.

At the back of the cottage there was a piece of waste ground extending to the river, with a small, old, ruinous barn standing on it a few yards from the bank. Between the barn and the stream the ground was overgrown with rank weeds, and here one day Caleb came by chance upon his cat eating something among the weeds—a good-sized, fresh-caught trout! On examining the ground he found it littered with the heads, fins, and portions of backbones of the trout their cat had been feeding on every day since they had been in possession of him. They did not destroy their favourite, nor tell anyone of their discovery, but they watched him and found that it was his habit to bring a trout every day to that spot, but how he caught his fish was never known.

Eventually their cat came to a tragic end, as all Wylye anglers will be pleased to hear. He was found on the railway track, at the spot where he had the habit of sitting, crushed as flat as a pancake. It was thought that while sitting on the rail in his usual way he had become so absorbed at the sight of the straight, shining line that the noise and vibration of the approaching train failed to arouse him in time to save himself. It seemed strange to them that a creature so very much alive and quick to escape danger should have met its death in this way; and what added to the wonder was that another cat of the village was found on the line crushed by a train shortly afterwards. Probably the sight of the shining rail gazed at too long and fixedly had produced a hypnotic effect on the animal's brain and made it powerless to escape.

It is rather an odd coincidence that in the village inn where I am writing a portion of this book, including the present chapter, there should be three cats, unlike one another in appearance and habits as three animals of different and widely separated species, one of them with a great resemblance to the shepherd's picture of his Doveton animal. All three were strays, which the landlady, who has a tender heart, took in when they were starving, and made pets of; and all are beautiful. One has Persian blood in him; a long-haired, black and brown animal with gold-coloured eyes; playful as a kitten, incessantly active, fond of going for a walk with some inmate of the house, and when no one—cat or human being—will have any more of him you will see him in the garden stalking a fly, or lying on his back on the ground under a beech-tree striking with his claws or catching at something invisible in the air—motes in the sunbeam. The second is a large, black cat with white collar and muzzle and sea-green eyes, and is of an indolent, luxurious disposition, lying coiled up by the hour on the most comfortable cushion it can find. The last is Gip, a magnificent creature, a third bigger than an average-sized cat—as large and powerfully built as the British wild cat, a tabby with opaline eyes, which show a pale green colour in some lights. These singular eyes, when I first saw this animal, almost startled me with their wild savage expression; nor was it a mere deceptive appearance, as I soon found. I never looked at this animal without finding these panther or lynx eyes fixed with a fierce intensity on me, and no sooner would I look towards him than he would crouch down, flatten his ears, and continue to watch my every movement as if apprehending a sudden attack on his life. It was many days before he allowed me to come near him without bounding away and vanishing, and not for two or three weeks would he suffer me to put a hand on him.

But the native wildness and suspicion in him could never be wholly overcome; it continued to show itself on occasions even after I had known him for months, and had won his confidence, and when it seemed that, in his wild cat, conditional way, he had accepted my friendship. He became lame, having injured one of his forelegs while hunting, and as the weather was cold he was pleased to spend his inactive and suffering time on the hearth-rug in my sitting-room. I found that rubbing warm, melted butter on his injured leg appeared to give him relief, and after the massaging and buttering he would lick the leg vigorously for ten or fifteen minutes, occasionally purring with satisfaction. Yet if I made any sudden movement, or rustled the paper in my hand, he would instantly spring up from his cushion at my feet and dart away to the door to make his escape; then, finding the door closed, he would sit down, recover his domesticity and return to my feet.

Yet this cat had been taken in as a kitten and had already lived some two or three years in the house, seeing many people and fed regularly with the others every day. It is not, however, very rare to meet with a cat of this disposition—the cat pure and simple, as nature made it, without that little tameness on the surface, or veneer of domesticity which life with man laid on it.

Gip was the most inveterate rat-killer I have ever known. He is never seen hunting other creatures, not even mice, although it is probable that he does kill and devour them on the spot, but he has the habit of bringing in the rats he captures; and as a day seldom passes on which he is not seen with one, and as sometimes two or three are brought in during the night, he cannot destroy fewer than three hundred to four hundred rats in the year.

Let anyone who knows the destructive powers of the rat consider what that means in an agricultural village, and what an advantage it is to the farmers to have their rick-yards and barns policed day and night by such an animal. For the whole village is his hunting-ground. His owner says that he is 'worth his weight in gold'. I should say that he is worth much more; that the equivalent in cash of his weight in purest gold, though he is big and heavy, would not be more than the value of the grain and other food-stuffs he saves from destruction in a single year.

He invariably brings in his rats alive to release and play with them in an old, stone-paved yard, and after a little play, he kills them, and if they are full-grown he leaves them; but when young he devours them or allows the other cats to have them.

Gip has a somewhat remarkable history. The village has always been a favourite resort of gipsies, and it happened that once when a party of gipsies had gone away it was discovered that they had left a litter of six kittens behind, and it greatly troubled the village mind to know what was to be done with them. 'Why didn't we drown them? Oh, no, we couldn't do that,' they said, 'they were several weeks old and past the time for drowning.' However, some one who kept ferrets turned up and kindly said he would take them to give to his ferrets, and everybody was satisfied to have the matter disposed of in that way. But it was a rather disgusting way, for although it seems quite natural to give little, living rodents to ferrets to be sucked of their warm blood, it goes against one's feelings to cast young cats to the pink-eyed beast; for the cat is a carnivorous creature, too, and not only so but is infinitely more beautiful and intelligent than the ferret and higher in the organic scale. However, these ideas did not prevail in the village, and the six kittens were taken; but they proved to be exceedingly vigorous and fierce for kittens, as if they knew what was going to be done to them: they fought and scratched, and eventually two, the biggest and fiercest of them, succeeded in making their escape, and by and by one of them was found on the premises at the inn—a refuge for all creatures in distress—which thereupon became its home.

Gipsies, I was told, are fond of keeping cats, and their cats are supposed to be the best ratters. As this was news to me I inquired of some gipsies in the neighbourhood, and they told me that it was true—they loved cats. They love lurchers, too, and no doubt they find cats with a genius for hunting very profitable pets. But how curiously varied hunting cats are in their tastes! You seldom find two quite alike. I have described one that was an accomplished trout-catcher, in spite of the ancient Gaelic proverb and universal saying that the cat loves fish but fears to wet its feet; also another who is exclusively a rat-killer. And here I recall an old story of a cat (an immortal puss) who only hunted pigeons. This tells that Sir Henry Wyatt was imprisoned in the Tower of London by Richard III, and was cruelly treated, having no bed to sleep on in his cell and scarcely food enough to keep him alive. One winter night, when he was half dead with cold, a cat appeared in his cell, having come down the chimney, and was very friendly, and slept curled up on his chest, thus keeping him warm all night. In the morning it vanished up the chimney, but appeared later with a pigeon, which it gave to Sir Henry, and then again departed. When the jailer appeared and repeated that he durst not bring more than the few morsels of food provided, Sir Henry then asked, 'Wilt thou dress any I provide?' This the jailer promised to do, for he pitied his prisoner, and taking the pigeon had it dressed and cooked for him. The cat continued bringing pigeons every day, and the jailer, thinking they were sent miraculously, continued to cook them, so that Sir Henry fared well, despite the order which Richard gave later, that no food at all was to be provided. He was getting impatient of his prisoner's power to keep alive on very little food, and he didn't want to behead him—he wanted him to die naturally. Thus in the end Sir Henry outlived the tyrant and was set free, and the family preserve the story to this day. It is classed as folk-lore, but there is no reason to prevent one from accepting it as literal truth.

It is a well-known habit of some hunting, or poaching, cats to bring their captives to their master or mistress. I have met with scores—I might say with hundreds—of such cases. I remember an old gaucho, a neighbour of mine in South America, who used to boast that he usually had a spotted tinamou—the partridge of the pampas—for his dinner every day, brought in by his cat. Even in England, where partridges are not so abundant or easily taken, there are clever partridge-hunting cats. I remember one, a very fine white cat, owned by a woodman I once lodged with in Savernake Forest, who was in the habit of bringing in a partridge and would place it on the kitchen floor and keep guard over it until the woodman's wife came to take it up and put it away for the Sunday's dinner. A lady friend told me of a cat at a farm-house where she was staying during the summer months, which became attached to her and was constantly bringing her young rabbits. They were never injured but held firmly by the skin of the neck. The lady would take the rabbit gently into her hands and deposit it on her lap, and cover it over with a handkerchief or a cloth, and pussy, seeing it safe in her power, would then go away. The lady would then walk away to a distance from the house to liberate the little trembler, devoutly wishing that this too affectionate cat would get over the delusion that such gifts were acceptable to her. One day pussy came trotting into the drawing-room with a stoat in his mouth, and depositing it on the carpet by the side of her chair immediately turned round and hastily left the room. The stoat was dead, not being a creature that could easily be carried about alive, and pussy, having other matters to attend to, did not think it necessary to wait to see her present taken up and carefully deposited in her mistress's lap.

Cases of this kind are exceedingly common, and the simple explanation is, that the cat is not quite so unsocial a creature as some naturalists would have us believe. We may say that in this respect he compares badly with elephants, whales, pigs, seals, cattle, apes, wolves, dogs, and other large-brained social mammals; but he does not live wholly for himself. He is able to take thought for other cats and for his human companion—master hardly seems the right word in the case of such an animal—who is doubtless to him only a very big cat that walks erect on his hind legs. I must, however, relate one more instance of a cat who hunted for others, told to me by a very aged friend of mine, a native of Fonthill Bishop, and some of whose early memories will be given later in the chapter entitled 'Old Wiltshire Days'.

When she was a young motherless girl and they were very poor indeed, her father being incapacitated, they had a cat that was a great help to them; a large black and white animal who spent a greater part of his time hunting in Fonthill Abbey woods, and who was always bringing in something for the pot. The cat was attached to her, and whatever was brought in was for her exclusively, and I imagine it is so in all cases in which a cat has the custom of bringing anything it catches into the house. The cat mind cannot understand a division of food. It does not and cannot share a mouse or bird with another cat, and when it gives it gives the whole animal, and to one person alone. When the cat brought a rabbit home he would not come into the kitchen with it if he saw her little brother or any other person there, lest they should take it into their hands; he would steal off and conceal it among the weeds at the back of the cottage, then come back to make little mewing sounds understood by its young mistress, and she would thereupon follow it out to where the rabbit was hidden and take it up, and the cat would then be satisfied.

The cat brought her rabbits and di-dappers, as she called the moorhen, caught in the sedges by the lake in the park. This was the first occasion of my meeting with this name for a bird, but it comes no doubt from dive-dapper, an old English vernacular name (found in Shakespeare) of the dabchick, or little grebe. Moorhens were not the only birds it captured: on two or three occasions it brought in a partridge and on one occasion a fish. Whether it was a trout or not she could not say; she only knew that she cooked and ate it and that it was very good.

One day, looking out, she spied her cat coming home with something very big, something it had caught larger than itself; and it was holding its head very high, dragging its burden along with great labour. It was a hare, and she ran out to receive it, and when she got to the cat and stooped down to take it from him he released it too soon, for it was uninjured, and away it bounded and vanished into the woods, leaving them both very much astonished and disgusted.

For a long time her cat managed to escape the poaching animal's usual fate in the woods, which were strictly preserved then, in the famous Squire Beckford's day, as they are now; but a day arrived when it came hobbling in with a broken leg. It had been caught in a steel trap, and some person who was not a keeper had found and released it. She washed the blood off, and taking it on her lap put the bones together and bound up the broken limb as well as she was able, and the bones joined, and before very long the cat was well again. And no sooner was it well than it resumed its hunting in the woods and bringing in rabbits and di-dappers.

But alas, it had but nine lives, and having generously spent all in the service of its young mistress it came to its end; at all events it finally disappeared, and it was conjectured that a keeper had succeeded in killing it.

One of my old shepherd's stories about strange or eccentric persons he had known during his long life was of a gentleman farmer, an old bachelor, in the parish of Winterbourne Bishop, who had (for a man) an excessive fondness for cats and who always kept eleven of these animals as pets. For some mysterious reason that number was religiously adhered to. The farmer was fond of riding on the downs, and was invariably attended by a groom in livery—a crusty old fellow; and one of this man's duties was to attend to his master's eleven cats. They had to be fed at their proper time, in their own dining-room, eating their meals from a row of eleven plates on a long, low table made expressly for them. They were taught to go each one to his own place and plate, and not to get on to the table, but to eat 'like Christians,' without quarrelling or interfering with their neighbours on either side. And, as a rule, they all behaved properly, except one big tom-cat, who developed so greedy, spiteful, and tyrannical a disposition that there was never a meal but he upset the harmony and brought it to a disorderly end, with spittings, snarlings, and scratchings. Day after day the old groom went to his master with a long, dolorous plaint of this cat's intolerable behaviour, but the farmer would not consent to its removal, or to any strong measures being taken; kindness combined with patience and firmness, he maintained, would at last win even this troublesome animal to a better mind. But in the end he, too, grew tired of this incorrigible cat, who was now making the others spiteful and quarrelsome by his example; and one day, hearing a worse account than usual, he got into a passion and taking a loaded gun handed it to the groom with orders to shoot the cat on the spot on the very next occasion of it misbehaving, so that not only would they be rid of it but its death in that way would serve as a warning to the others. At the very next meal the bad cat got up the usual row, and by and by they were all fighting and tearing each other on the table, and the groom, seizing the gun, sent a charge of shot into the thickest of the fight, shooting three of the cats dead. But the author of all the mischief escaped without so much as a pellet! The farmer was in a great rage at this disastrous blundering, and gave notice to his groom on the spot; but the man was an old and valued servant, and by and by he forgave him, and the quarrelsome animal having been got rid of, and four fresh cats obtained to fill up the gaps, peace was restored.

I must now return to the subject of the cat tragedy related in the early part of this chapter—Bawcombe's cat at Doveton, who had the habit of sitting on a rail of the line which runs through the vale, and was eventually killed by a train. So strange a story—for how strange it seems that an animal of so cautious and well-balanced a mind, so capable above all others of saving itself in difficult and dangerous emergencies, should have met its end in such a way!—might very well have suggested something behind the mere fact, some mysterious weakness in the animal similar to that which Herodotus relates of the Egyptian cat in its propensity of rushing into the fire when a house was burning and thus destroying itself. Yet no such idea came into my mind: it was just a 'strange fact,' an accident in the life of an individual, and after telling it I passed on, thinking no more about the subject, only to find long months afterwards, by the merest chance, that I had been very near to a discovery of the greatest significance and interest in the life-history of the animal.

It came about in the following way. I was on the platform of a station on the South-Western line from Salisbury to Yeovil, waiting for my train, when a pretty little kitten came out of the stationmaster's house at the end of the platform, and I picked it up. Then a child, a wee girlie of about five, came out to claim her pet, and we got into a talk about the kitten. She was pleased at its being admired, and saying she would show me the other one, ran in and came out with a black kitten in her arms. I duly admired this one too. 'But,' I said, 'they won't let you keep both, because then there would be too many cats.'

'No—only two,' she returned.

'Three, with their mother.'

'No, they haven't got a mother—she was killed on the line.'

I remembered the shepherd's cat, and by and by finding the stationmaster I questioned him about the cat that had been killed.

'Oh, yes,' he said, 'cats are always getting killed on the line—we can never keep one long. I don't know if they try to cross the line or how it is. One of the porters saw the last one get killed and will tell you just how it happened.'

I found the porter, and his account was that he saw the cat on the line, standing with its forepaws on a rail when an express train was coming. He called to the cat two or three times, then yelled at it to frighten it off, but it never moved; it stared as if dazed at the coming train, and was struck on the head and knocked dead.

This story set me making enquiries at other village stations, and at other villages where there are no stations, but close to which the line runs in the Wylye vale and where their is a pointsman. I was told that cats are very often found killed on the line, in some instances crushed as if they had been lying or sitting on a rail when the train went over them. They get dazed, the men said, and could not save themselves.

I was also told that rabbits were sometimes killed and, more frequently, hares. 'I've had several hares from the line,' one man told me. He said that he had seen a hare running before a train and thought that in most cases the hare kept straight on until it was run down and killed. But not in every case, as he had actually seen one hare killed, and in this case the hare sat up and remained staring at the coming train until it was struck.

It cannot be doubted, I think, that the cat is subject to this strange weakness. It is not a case of 'losing its head' like a cyclist amidst the traffic in a thoroughfare, or of miscalculating the speed of a coming train and attempting too late to cross the line. The sight of the coming train paralyses its will, or hypnotizes it, and it cannot save itself.

Now the dog, a less well-balanced animal than the cat and inferior in many ways, has no such failing and is killed by a train purely through blundering. While engaged in making these inquiries, a Wiltshire woman told me of an adventure she had with her dog, a fox terrier. She had just got over the line at a level crossing when the gates swung to, and looking for her dog she saw him absorbed in a smell he had discovered on the other side of the line. An express train was just coming, and screaming to her dog she saw him make a dash to get across just as the engine came abreast of her. The dog had vanished from sight, but when the whole train had passed up he jumped from between the rails where he had been crouching and bounded across to her, quite unhurt. He had dived under the train behind the engine, and waited there till it had gone by!

It is, however, a fact that not all the cats killed on the line have been hypnotized or dazed at the sight of the coming train; undoubtedly some do meet their death through attempting to cross the line before a coming train. At all events, I heard of one such case from a person who had witnessed it. It was at a spot where a small group of workmen's cottages stands close to the line at a village; here I was told that 'several cats' got killed on the line every year, and as the man who gave me the information had seen a cat running across the line before a train and getting killed it was assumed by the cottagers that it was so in all cases.

XVI
THE ELLERBYS OF DOVETON

The Bawcombes at Doveton Farm—Caleb finds favour with his master—
Mrs. Ellerby and the shepherd's wife—The passion of a childless wife—
The curse—A story of the 'mob'—The attack on the farm—A man trans-
ported for life—The hundred and ninth Psalm—The end of the Ellerbys

CALEB and his wife invariably spoke of their time at Doveton Farm
in a way which gave one the idea that they regarded it as the most
important period of their lives. It had deeply impressed them, and
doubtless it was a great change for them to leave their native village for
the first time in their lives and go long miles from home among strangers to serve a new master. Above everything they felt leaving the old
father who was angry with them, and had gone to the length of dis-
owning them for taking such a step. But there was something besides
all this which had served to give Doveton an enduring place in their
memories, and after many talks with the old couple about their Warminster days I formed the idea that it was more to them than any
other place where they had lived, because of a personal feeling they
cherished for their master and mistress there.

Hitherto Caleb had been in the service of men who were but a little
way removed in thought and feeling from those employed. They were
mostly small men, born and bred in the parish, some wholly self-made,
with no interest or knowledge of anything outside their own affairs,
and almost as far removed as the labourers themselves from the ranks
above. The Ellerbys were of another stamp, or a different class. If not a
gentleman, Mr. Ellerby was very like one and was accustomed to asso-
ciate with gentlemen. He was a farmer, descended from a long line of
farmers; but he owned his own land, and was an educated and travelled man, considered wealthy for a farmer; at all events he was able
to keep his carriage and riding and hunting horses in his stables, and
he was regarded as the best breeder of sheep in the district. He lived in
a good house, which with its pictures and books and beautiful deco-
rations and furniture appeared to their simple minds extremely luxurious. This atmosphere was somewhat disconcerting to them at first,
for although he knew his own value, priding himself on being a 'good

shepherd,' Caleb had up till now served with farmers who were in a sense on an equality with him, and they understood him and he them. But in a short time the feeling of strangeness vanished: personally, as a fellow-man, his master soon grew to be more to him than any farmer he had yet been with. And he saw a good deal of his master. Mr. Ellerby cultivated his acquaintance, and, as we have seen, got into the habit of seeking him out and talking to him even when he was at a distance out on the down with his flock. And Caleb could not but see that in this respect he was preferred above the other men employed on the farm—that he had 'found favour' in his master's eyes.

When he had told me that story about Watch and the corn-crake, it stuck in my mind, and on the first opportunity I went back to that subject to ask what it really was that made his master act in such an extraordinary manner—to go out on a pouring wet day in a summer suit and straw hat, and walk a mile or two just to stand there in the rain talking to him about nothing in particular. What secret trouble had he—was it that his affairs were in a bad way, or was he quarrelling with his wife? No, nothing of the kind, it was a long story—this secret trouble of the Ellerbys, and with his unconquerable reticence in regard to other people's private affairs he would have passed it off with a few general remarks.

But there was his old wife listening to us, and, woman-like, eager to discuss such a subject, she would not let it pass. She would tell it and would not be silenced by him: they were all dead and gone—why should I not be told if I wanted to hear it? And so with a word put in here and there by him when she talked, and with a good many words interposed by her when he took up the tale, they unfolded the story, which was very long as they told it and must be given briefly here.

It happened that when the Bawcombes settled at Doveton, just as Mr. Ellerby had taken to the shepherd, making a friend of him, so Mrs. Ellerby took to the shepherd's wife, and fell into the habit of paying frequent visits to her in her cottage. She was a very handsome woman, of a somewhat stately presence, dignified in manner, and she wore her abundant hair in curls hanging on each side to her shoulders—a fashion common at that time. From the first she appeared to take a particular interest in the Bawcombes, and they could not but notice that she was more gracious and friendly towards them than to the others of their station on the farm. The Bawcombes had three children then, aged 6, 4, and 2 years respectively, all remarkably healthy, with rosy cheeks and black eyes, and they were merry-tempered little things. Mrs. Ellerby appeared much taken with the children; praised their mother for always keeping them to clean and nicely dressed, and wondered how she could manage it on their small earnings. The carter and his wife lived in a cottage close by, and they, too, had three little children, and next to the carter's was the bailiff's cottage, and he, too, was married and had children; but Mrs. Ellerby never went into their cottages, and the shepherd and his wife concluded that it was because in both cases the children were rather puny, sickly-looking little things and were never very clean. The carter's wife, too, was a slatternly woman. One day when Mrs. Ellerby came in to see Mrs. Bawcombe the carter's wife was just going out of the door, and Mrs. Ellerby appeared displeased, and before leaving she said, 'I hope, Mrs. Bawcombe, you are not going to mix too freely with your neighbours or let your children go too much with them and fall into their ways.' They also observed that when she passed their neighbours' children in the lane she spoke no word and appeared not to see them. Yet she was kind to them too, and whenever she brought a big parcel of cakes, fruit, and sweets for the children, which she often did, she would tell the shepherd's wife to divide it into three lots, one for her own children and the others for those of her two neighbours. It was clear to see that Mrs. Ellerby had grown fond of her children, especially of the eldest, the little rosy-cheeked six-year-old boy. Sitting in the cottage she would call him to her side and would hold his hand while conversing with his mother; she would also bare the child's arm just for the pleasure of rubbing it with her hand and clasping it round with her fingers, and sometimes when caressing the child in this way she would turn her face aside to hide the tears that dropped from her eyes.

She had no child of her own—the one happiness which she and her husband desired above all things. Six times in their ten married years they had hoped and rejoiced, although with fear and trembling, that their prayer would be answered, but in vain—every child born to them came lifeless into the world. 'And so 'twould always be, for sure,' said the villagers, 'because of the curse.'

For it was a cause of wonder to the shepherd and his wife that this couple, so strong and healthy, so noble-looking, so anxious to have children, should have been so unfortunate, and still the villagers repeated that it was the curse that was on them.

This made the shepherd angry. 'What be you saying about a curse that is on them?—a good man and a good woman!' he would exclaim, and taking up his crook go out and leave them to their gossip. He would not ask them what they meant; he refused to listen when they tried to tell him; but in the end he could not help knowing, since the idea had become a fixed one in the minds of all the villagers, and he could not keep it out. 'Look at them,' the gossipers would say, 'as fine a couple as you ever saw, and no child; and look at his two brothers, fine, big, strong, well-set up men, both married to fine healthy women, and never a child living to any of them. And the sisters unmarried! 'Tis the curse and nothing else.'

The curse had been uttered against Mr. Ellerby's father, who was in his prime in the year 1831 at the time of the 'mob,' when the introduction of labour-saving machinery in agriculture sent the poor farm-labourers mad all over England. Wheat was at a high price at that time, and the farmers were exceedingly prosperous, but they paid no more than seven shillings a week to their miserable labourers. And if they were half-starved when there was work for all, when the corn was reaped with sickles, what would their condition be when reaping machines and other new implements of husbandry came into use? They would not suffer it; they would gather in bands everywhere and destroy the machinery, and being united they would be irresistible; and so it came about that there were risings or 'mobs' all over the land.

Mr. Ellerby, the most prosperous and enterprising farmer in the parish, had been the first to introduce the new methods. He did not believe that the people would rise against him, for he well knew that he was regarded as a just and kind man and was even loved by his own labourers, but even if it had not been so he would not have hesitated to carry out his resolution, as he was a high-spirited man. But one day the villagers got together and came unexpectedly to his barns, where they set to work to destroy his new thrashing machine. When he was told he rushed out and went in hot haste to the scene, and as he drew near some person in the crowd threw a heavy hammer at him, which struck him on the head and brought him senseless to the ground.

He was not seriously injured, but when he recovered the work of destruction had been done and the men had gone back to their homes, and no one could say who had led them and who had thrown the hammer. But by and by the police discovered that the hammer was the property of a shoemaker in the village, and he was arrested and charged with injuring with intent to murder. Tried with many others from other villages in the district at the Salisbury Assizes, he was found guilty and sentenced to transportation for life. Yet the Doveton shoe-maker was known to every one as a quiet, inoffensive young man, and to the last he protested his innocence, for although he had gone with the others to the farm he had not taken the hammer and was guiltless of having thrown it.

Two years after he had been sent away Mr. Ellerby received a letter with an Australian postmark on it, but on opening it found nothing but a long denunciatory passage from the Bible enclosed, with no name or address. Mr. Ellerby was much disturbed in his mind, and instead of burning the paper and holding his peace, he kept it and spoke about it to this person and that, and every one went to his Bible to find out what message the poor shoemaker had sent, for it had been discovered that it was the one hundred and ninth Psalm, or a great portion of it, and this is what they read:

'Hold not Thy peace, O God of my praise: for the mouth of the wicked and the mouth of the deceitful are opened against me: they have spoken against me with a lying tongue. They compassed me about also with words of hatred; and fought against me without a cause. And they have rewarded me evil for good and hatred for my love.

'Set Thou a wicked man over him; and let Satan stand at his right hand.

'When he shall be judged, let him be condemned; and let his prayer become sin.

'Let his days be few; and another take his office.

'Let his children be fatherless, and his wife a widow.

'Let his children be continually vagabonds, and beg; let them seek their bread also out of their desolate places.

'Let there be none to extend mercy unto him; neither let there be any to favour his fatherless children.

'Let his posterity be cut off; and in the generation following let their name be blotted out.

'Let the iniquity of his fathers be remembered with the Lord; and let not the sin of his mother be blotted out.

'Let them be before the Lord continually, that he may cut off the memory of them from the earth.

'Because that he remembered not to show mercy, but persecuted the poor and needy man, that he might even slay the broken in heart.

'As he loved cursing, so let it come unto him; as he delighted not in blessing, so let it be far from him.

'As he clothed himself with cursing like as with a garment, so let it come into his bowels like water, and like oil into his bones.

'Let it be unto him as a garment which covereth him, and for a girdle wherewith he is girded continually.

'But do Thou for me, O God the Lord, for Thy name's sake. For I am poor and needy, and my heart is wounded within me.

'I am gone like the shadow when it declineth : I am tossed up and down as the locust.

'My knees are weak through fasting; and my flesh faileth of fatness.

'Help me, O Lord my God; that they may know that this is Thy hand; that Thou, Lord, hast done it.'

From that time the hundred and ninth Psalm became familiar to the villagers, and there were probably not many who did not get it by heart. There was no doubt in their minds of the poor shoemaker's innocence. Every one knew that he was incapable of hurting a fly. The crowd had gone into his shop and swept him away with them—all were in it; and some person seeing the hammer had taken it to help in smashing the machinery. And Mr. Ellerby had known in his heart that he was innocent, and if he had spoken a word for him in court he would have got the benefit of the doubt and been discharged. But no, he wanted to have his revenge on some one, and he held his peace and allowed this poor fellow to be made the victim. Then, when he died, and his eldest son succeeded him at Doveton Farm, and he and the other sons got married, and there were no children, or none born alive, they went back to the Psalm again and read and re-read and quoted the words: 'Let his posterity be cut off; and in the generation following let their name be blotted out.' Undoubtedly the curse was on them!

Alas! it was; the curse was their belief in the curse, and the dreadful effect of the knowledge of it on a woman's mind—all the result of Mr. Ellerby the father's fatal mistake in not having thrown the scrap of paper that came to him from the other side of the world into the fire. All the unhappiness of the 'generation following' came about in this way, and the family came to an end; for when the last of the Ellerbys died at a great age there was not one person of the name left in that part of Wiltshire.

XVII

OLD WILTSHIRE DAYS

Old memories—Hindon as a borough and as a village—The Lamb Inn and its birds—The 'mob' at Hindon—The blind smuggler—Rawlings of Lower Pertwood Farm—Reed, the thresher and deer-stealer—He leaves a fortune —Devotion to work—Old Father Time—Groveley Wood and the people's rights—Grace Reed and the Earl of Pembroke—An illusion of the very aged—Sedan-chairs in Bath—Stick-gathering by the poor—Game-preserving

THE incident of the unhappy young man who was transported to Australia or Tasmania, which came out in the shepherd's history of the Ellerby family, put it in my mind to look up some of the very aged people of the downland villages, whose memories could go back to the events of eighty years ago. I found a few, 'still lingering here,' who were able to recall that miserable and memorable year of 1830 and had witnessed the doings of the 'mobs.' One was a woman, my old friend of Fonthill Bishop, now aged 94, who was in her teens when the poor labourers 'a thousand strong,' some say, armed with cudgels, hammers and axes, visited her village and broke up the threshing machines they found there.

Another person who remembered that time was an old but remarkably well-preserved man of 89 at Hindon, a village a couple of miles distant from Fonthill Bishop. Hindon is a delightful little village, so rustic and pretty amidst its green, swelling downs, with great woods crowning the heights beyond, that one can hardly credit the fact that it was formerly an important market and session town and a Parliamentary borough returning two members; also that it boasted among other greatnesses thirteen public-houses. Now it has two, and not flourishing in these tea- and mineral-water drinking days. Naturally it was an exceedingly corrupt little borough, where free beer for all was the order of the day for a period of four to six weeks before an election, and where every householder with a vote looked to receive twenty guineas from the candidate of his choice. It is still remembered that when a householder in those days was very hard up, owing, perhaps, to his too frequent visits to the thirteen public-houses, he would go to some substantial tradesman in the place and *pledge* his twenty guineas, due at the next election! In due time, after the Reform Bill, it was deprived of its glory, and later when the South-Western Railway built their line from Salisbury to Yeovil and left Hindon some miles away, making their station at Tisbury, it fell into decay, dwindling to the small village it now is; and its last state, sober and purified, is very much better than the old. For although sober, it is contented and even merry, and exhibits such a sweet friendliness towards the stranger within its gates as to make him remember it with pleasure and gratitude.

What a quiet little place Hindon has become, after its old noisy period, the following little bird story will show. For several weeks during the spring and summer of 1909 my home was at the Lamb Inn, a famous posting-house of the great old days, and we had three pairs of birds—throstle, pied wagtail, and flycatcher—breeding in the ivy covering the wall facing the village street, just over my window. I watched them when building, incubating, feeding their young, and bringing their young off. The villagers, too, were interested in the sight, and sometimes a dozen or more men and boys would gather and stand for half an hour watching the birds flying in and out of their nests when feeding their young. The last to come off were the flycatchers, on 18 June. It was on the morning of the day I left, and one of the little things flitted into the room where I was having my breakfast. I succeeded in capturing it before the cats found out, and put it back on the ivy. There were three young birds; I had watched them from the time they hatched, and when I returned a fortnight later, there were the three, still being fed by their parents in the trees and on the roof, their favourite perching-place being on the swinging sign of the 'Lamb.' Whenever an old bird darted at and captured a fly the three young would flutter round it like three butterflies to get the fly. This continued until 18 July, after which date I could not detect their feeding the young, although the hunger-call was occasionally heard.

If the flycatcher takes a month to teach its young to catch their own flies, it is not strange that it breeds but once in the year. It is a delicate art the bird practises and takes long to learn, but how different with the martin, which dismisses its young in a few days and begins breeding again, even to the third time!

These three broods over my window were not the only ones in the place; there were at least twenty other pairs in the garden and outhouses of the inn—sparrows, thrushes, blackbirds, dunnocks, wrens, starlings, and swallows. Yet the inn was in the very centre of the village, and being an inn was the most frequented and noisiest spot.

To return to my old friend of 89. He was but a small boy, attending the Hindon school, when the rioters appeared on the scene, and he watched their entry from the schoolhouse window. It was market-day, and the market was stopped by the invaders, and the agricultural machines brought for sale and exhibition were broken up. The picture that remains in his mind is of a great excited crowd in which men and cattle and sheep were mixed together in the wide street, which was the market-place, and of shouting and noise of smashing machinery, and finally of the mob pouring forth over the down on its way to the next village, he and other little boys following their march.

The smuggling trade flourished greatly at that period, and there were receivers and distributors of smuggled wine, spirits, and other commodities in every town and in very many villages throughout the county in spite of its distance from the sea-coast. One of his memories is of a blind man of the village, or town as it was then, who was used as an assistant in this business. He had lost his sight in childhood, one eye having been destroyed by a ferret which got into his cradle; then, when he was about six years old, he was running across the room one day with a fork in his hand when he stumbled, and falling on the floor had the other eye pierced by the prongs. But in spite of his blindness he became a good worker, and could make a fence, reap, trim hedges, feed the animals, and drive a horse as well as any man. His father had a small farm and was a carrier as well, a quiet, sober, industrious man who was never suspected by his neighbours of being a smuggler, for he never left his house and work, but from time to time he had little consignments of rum and brandy in casks received on a dark night and carefully stowed away in his manure heap and in a pit under the floor of his pigsty. Then the blind son would drive his old mother in the carrier's cart to Bath and call at a dozen or twenty private houses, leaving parcels which had been already ordered and paid for—a gallon of brandy in one, two or four gallons of rum at another, and so on, until all was got rid of, and on the following day they would return with goods to Hindon. This quiet little business went on satisfactorily for some years, during which the officers of the excise had stared a thousand times with their eagle's eyes at the quaint old woman in her poke bonnet and shawl, driven by a blind man with a vacant face, and had suspected nothing, when a little mistake was made and a jar of brandy delivered at a wrong address. The recipient was an honest gentleman, and in his anxiety to find the rightful owner of the brandy made extensive inquiries in his neighbourhood, and eventually the excisemen got wind of the affair, and on the very next visit of the old woman and her son to Bath they were captured. After an examination before a magistrate the son was discharged on account of his blindness, but the cart and horses, as well as the smuggled spirits, were confiscated, and the poor blind man had to make his way on foot to Hindon.

Another of his recollections is of a family named Rawlings, tenants of Lower Pertwood Farm, near Hindon, a lonely, desolate-looking house hidden away in a deep hollow among the high downs. The Farmer Rawlings of seventy to eighty years ago was a man of singular ideas, and that he was permitted to put them in practice shows that severe as was the law in those days, and dreadful the punishments inflicted on offenders, there was a kind of liberty which does not exist now—the liberty a man had of doing just what he thought proper in his own house. This Rawlings had a numerous family, and some died at home and others lived to grow up and go out into the world under strange names—Faith, Hope, and Charity were three of his daughters, and Justice, Morality, and Fortitude three of his sons. Now, for some reason Rawlings objected to the burial of his dead in the churchyard of the nearest village—Monkton Deverill, and the story is that he quarrelled with the rector over the question of the church bell being tolled for the funeral. He would have no bell tolled, he swore, and the rector would bury no one without the bell. Thereupon Rawlings had the coffined corpse deposited on a table in an outhouse and the door made fast. Later there was another death, then a third, and all three were kept in the same place for several years, and although it was known to the whole countryside no action was taken by the local authorities.

My old informant says that he was often at the farm when he was a young man, and he used to steal round to the 'Dead House,' as it was called, to peep through a crack in the door and see the three coffins resting on the table in the dim interior.

Eventually the dead disappeared a little while before the Rawlings gave up the farm, and it was supposed that the old farmer had buried them in the night-time in one of the neighbouring chalk-pits, but the spot has never been discovered.

One of the stories of the old Wiltshire days I picked up was from an old woman, aged 87, in the Wilton workhouse. She had a vivid recollection of a labourer named Reed, in Odstock, a village on the Ebble near Salisbury, a stern, silent man, who was a marvel of strength and endurance. The work in which he most delighted was precisely that which most labourers hated, before threshing machines came in despite the action of the 'mobs'—threshing out corn with the flail. From earliest dawn till after dark he would sit or stand in a dim, dusty barn, monotonously pounding away, without an interval to rest, and without dinner, and with no food but a piece of bread and a pinch of salt. Without the salt he would not eat the bread. An hour after all others had ceased from work he would put on his coat and trudge home to his wife and family.

The woman in the workhouse remembers that once, when Reed was a very old man past work, he came to their cottage for something, and while he stood waiting at the entrance, a little boy ran in and asked his mother for a piece of bread and butter with sugar on it. Old Reed glared at him, and shaking his big stick exlaimed 'I'd give you sugar with this if you were my boy!' and so terrible did he look in his anger at the luxury of the times, that the little boy burst out crying and ran away!

What chiefly interested me about this old man was that he was a deer-stealer of the days when that offence was common in the country. It was not so great a crime as sheep-stealing, for which men were hanged; taking a deer was punished with nothing worse than hard labour as a rule. But Reed was never caught; he would labour his full time and steal away after dark over the downs, to return in the small hours with a deer on his back. It was not for his own consumption; he wanted the money for which he sold it in Salisbury; and it is probable that he was in league with other poachers, as it is hard to believe that he could capture the animals single-handed.

After his death it was found that old Reed had left a hundred pounds to each of his two surviving daughters, and it was a wonder to everybody how he had managed not only to bring up a family and keep himself out of the workhouse to the end of his long life but to leave so large a sum of money. One can only suppose that he was a rigid economist and never had a week's illness, and that by abstaining from beer and tobacco he was able to save a couple of shillings each week out of his wages of seven or eight shillings; this, in forty years, would make the two hundred pounds with something over.

It is not a very rare thing to find a farm-labourer like old Reed of Odstock, with not only a strong preference for a particular kind of work, but a love of it as compelling as that of an artist for his art. Some friends of mine whom I went to visit over the border in Dorset told me of an enthusiast of this description who had recently died in the village. 'What a pity you did not come sooner,' they said. Alas! it is nearly always so; on first coming to stay at a village one is told that it has but just lost its oldest and most interesting inhabitant—a relic of the olden time.

This man had taken to the scythe as Reed had to the flail, and was never happy unless he had a field to mow. He was a very tall old man, so lean that he looked like a skeleton, the bones covered with a skin as brown as old leather, and he wore his thin grey hair and snow-white beard very long. He rode on a white donkey, and was usually seen mounted galloping down the village street, hatless, his old brown, bare feet and legs drawn up to keep them from the ground, his scythe over his shoulder. 'Here comes old Father Time,' they would cry, as they called him, and run to the door to gaze with ever fresh delight at the wonderful old man as he rushed by, kicking and shouting at his donkey to make him go faster. He was always in a hurry, hunting for work with furious zeal, and when he got a field to mow so eager was he that he would not sleep at home, even if it was close by, but would lie down on the grass at the side of the field and start working at dawn, between two and three o'clock, quite three hours before the world woke up to its daily toil.

The name of Reed, the zealous thresher with the flail, serves to remind me of yet another Reed, a woman who died a few years ago aged 94, and whose name should be cherished in one of the downland villages. She was a native of Barford St. Martin on the Nadder, one of two villages, the other being Wishford, on the Wylye river, the inhabitants of which have the right to go into Groveley Wood, an immense forest on the Wilton estate, to obtain wood for burning, each person being entitled to take home as much wood as he or she can carry. The people of Wishford take green wood, but those of Barford only dead, they having bartered their right at a remote period to cut growing trees for a yearly sum of five pounds, which the lord of the manor still pays to the village, and, in addition, the right to take dead wood.

It will be readily understood that this right possessed by the people of two villages, both situated within a mile of the forest, has been a perpetual source of annoyance to the noble owners in modern times, since the strict preservation of game, especially of pheasants, has grown to be almost a religion to the landowners. Now it came to pass that about half a century or longer ago, the Pembroke of that time made the happy discovery, as he imagined, that there was nothing to show that the Barford people had any right to the dead wood. They had been graciously allowed to take it, as was the case all over the country at that time, and that was all. At once he issued an edict prohibiting the taking of dead wood from the forest by the villagers, and great as the loss was to them they acquiesced; not a man of Barford St. Martin dared to disobey the prohibition or raise his voice against it. Grace Reed then determined to oppose the mighty earl, and accompanied by four other women of the village boldly went to the wood and gathered their sticks and brought them home. They were summoned before the magistrates and fined, and on their refusal to pay were sent to prison; but the very next day they were liberated and told that a mistake had been made, that the matter had been inquired into, and it had been found that the people of Barford did really have the right they had exercised so long to take dead wood from the forest.

As a result of the action of these women the right has not been challenged since, and on my last visit to Barford, a few days before writing this chapter, I saw three women coming down from the forest with as much dead wood as they could carry on their heads and backs. But how near they came to losing their right! It was a bold, an unheard-of thing which they did, and if there had not been a poor cottage woman with the spirit to do it at the proper moment the right could never have been revived.

Grace Reed's children's children are living at Barford now; they say that to the very end of her long life she preserved a very clear memory of the people and events of the village in the old days early in the last century. They say, too, that in recalling the far past, the old people and scenes would present themselves so vividly to her mind that she would speak of them as of recent things, and would say to some one fifty years younger than herself, 'Can't you remember it? Surely you haven't forgotten it when 'twas the talk of the village!'

It is a common illusion of the very aged, and I had an amusing instance of it in my old Hindon friend when he gave me his first impressions of Bath as he saw it about the year 1835. What astonished him most were the sedan-chairs, for he had never even heard of such a conveyance, but here in this city of wonders you met them in every street. Then he added, 'But you've been to Bath and of course you've seen them, and know all about it.'

About firewood-gathering by the poor in woods and forests, my old friend of Fonthill Bishop says that the people of the villages adjacent to the Fonthill and Great Ridge Woods were allowed to take as much dead wood as they wanted from those places. She was accustomed to go to the Great Ridge Wood, which was even wilder and more like a natural forest in those days than it is now. It was fully two miles from her village, a longish distance to carry a heavy load, and it was her custom after getting the wood out to bind it firmly in a large barrel-shaped bundle or faggot, as in that way she could roll it down the

smooth steep slopes of the down and so get her burden home without so much groaning and sweating. The great wood was then full of hazel-trees, and produced such an abundance of nuts that from mid-July to September people flocked to it for the nutting from all the country round, coming even from Bath and Bristol to load their carts with nuts in sacks for the market. Later, when the wood began to be more strictly preserved for sporting purposes, the rabbits were allowed to increase excessively, and during the hard winters they attacked the hazel-trees, gnawing off the bark, until this most useful and profitable wood the forest produced—the scrubby oaks having little value—was well nigh extirpated. By and by pheasants as well as rabbits were strictly preserved, and the firewood-gatherers were excluded altogether. At present you find dead wood lying about all over the place, abundantly as in any primitive forest, where trees die of old age or disease, or are blown down or broken off by the winds and are left to rot on the ground, overgrown with ivy and brambles. But of all this dead wood not a stick to boil a kettle may be taken by the neighbouring poor lest the pheasants should be disturbed or a rabbit be picked up.

Some more of the old dame's recollections will be given in the next chapter, showing what the condition of the people was in this district about the year 1830, when the poor farm-labourers were driven by hunger and misery to revolt against their masters—the farmers who were everywhere breaking up the downs with the plough to sow more and still more corn, who were growing very fat and paying higher and higher rents to their fat landlords, while the wretched men that drove the plough had hardly enough to satisfy their hunger.

XVIII

OLD WILTSHIRE DAYS—CONTINUED

An old Wiltshire woman's memories—Her home—Work on a farm—
A little bird-scarer—Housekeeping—The agricultural labourers' rising
—Villagers out of work—Relief work—A game of ball with barley
bannocks—Sheep-stealing—A poor man hanged—Temptations to steal
—A sheep-stealing shepherd—A sheep-stealing farmer—Story of Ebe-
nezer Garlick—A sheep-stealer at Chitterne—The law and the judges
—A 'human devil' in a black cap—How the revolting labourers were
punished—A last scene at Salisbury Court House—Inquest on a mur-
dered man—Policy of the farmers

THE STORY of her early life told by my old friend Joan, aged 94,
will serve to give some idea of the extreme poverty and hard suffering life of the agricultural labourers during the thirties of last century, at a time when farmers were exceedingly prosperous and landlords drawing high rents.

She was 3 years old when her mother died, after the birth of a
boy, the last of eleven children. There was a dame's school in their
little village of Fonthill Abbey, but the poverty of the family would
have made it impossible for Joan to attend had it not been for an un-
selfish person residing there, a Mr. King, who was anxious that every
child should be taught its letters. He paid for little Joan's schooling
from the age of 4 to 8; and now, in the evening of her life, when she
sits by the fire with her book, she blesses the memory of the man,
dead these seventy or eighty years, who made this solace possible for
her.

After the age of 8 there could be no more school, for now all
the older children had gone out into the world to make their own
poor living, the boys to work on distant farms, the girls to service or
to be wives, and Joan was wanted at home to keep house for her
father, to do the washing, mending, cleaning, cooking, and to be
mother to her little brother as well.

Her father was a ploughman, at seven shillings a week; but when
Joan was 10 he met with a dreadful accident when ploughing with a
couple of young or intractable oxen; in trying to stop them he got
entangled in the ropes and one of his legs badly broken by the plough.

As a result it was six months before he could leave his cottage. The overseer of the parish, a prosperous farmer who had a large farm a couple of miles away, came to inquire into the matter and see what was to be done. His decision was that the man would receive three shillings a week until able to start work again, and as that would just serve to keep him, the children must go out to work. Meanwhile, one of the married daughters had come to look after her father in the cottage, and that set the little ones free.

The overseer said he would give them work on his farm and pay them a few pence apiece and give them their meals; so to his farm they went, returning each evening home. That was her first place, and from that time on she was a toiler, indoors and out, but mainly in the fields, till she was past 85;—seventy-five years of hard work—then less and less as her wonderful strength diminished, and her sons and daughters were getting grey, until now at the age of 94 she does very little—practically nothing.

In that first place she had a very hard master in the farmer and overseer. He was known in all the neighbourhood as 'Devil Turner,' and even at that time, when farmers had their men under their heel as it were, he was noted for his savage tyrannical disposition; also for a curious sardonic humour, which displayed itself in the forms of punishment he inflicted on the workmen who had the ill-luck to offend him. The man had to take the punishment, however painful or disgraceful, without a murmur, or go and starve. Every morning thereafter Joan and her little brother, aged 7, had to be up in time to get to the farm at five o'clock in the morning, and if it was raining or snowing or bitterly cold, so much the worse for them, but they had to be there, for Devil Turner's bad temper was harder to bear than bad weather. Joan was a girl of all work, in and out of doors, and, in severe weather, when there was nothing else for her to do, she would be sent into the fields to gather flints, the coldest of all tasks for her little hands.

'But what could your little brother, a child of 7, do in such a place?' I asked.

She laughed when she told me of her little brother's very first day at the farm. The farmer was, for a devil, considerate, and gave him something very light for a beginning, which was to scare the birds from the ricks. 'And if they will come back you must catch them,' he said, and left the little fellow to obey the difficult command as he could. The birds that worried him most were the fowls, for however often he hunted them away they would come back again. Eventually, he found some string, with which he made some little loops fastened to sticks, and these he arranged on a spot of ground he had cleared, scattering a few grains of corn on it to attract the 'birds.' By this means he succeeded in capturing three of the robbers, and when the farmer came round at noon to see how he was getting on, the little fellow showed him his captures. 'These are not birds,' said the farmer, 'they are fowls, and don't you trouble yourself any more about them, but keep your eye on the sparrows and little birds and rooks and jackdaws that come to pull the straws out.'

That was how he started; then from the ricks to bird-scaring in the fields and to other tasks suited to one of his age, not without much suffering and many tears. The worst experience was the punishment of standing motionless for long hours at a time on a chair placed out in the yard, full in sight of the windows of the house, so that he could be seen by the inmates; the hardest, the cruellest task that could be imposed on him would come as a relief after this. Joan suffered no punishment of that kind; she was very anxious to please her master and worked hard; but she was an intelligent and spirited child, and as the sole result of her best efforts was that more and more work was put on her, she revolted against such injustice, and eventually, tried beyond endurance, she ran away home and refused to go back to the farm any more. She found some work in the village; for now her sister had to go back to her husband, and Joan had to take her place and look after her father and the house as well as earn something to supplement the three shillings a week they had to live on.

After about nine months her father was up and out again and went back to the plough; for just then a great deal of down was being broken up and brought under cultivation on account of the high price of wheat and good ploughmen were in request. He was lame, the injured limb being now considerably shorter than the other, and when ploughing he could only manage to keep on his legs by walking with the longer one in the furrow and the other on the higher ground. But after struggling on for some months in this way, suffering much pain and his strength declining, he met with a fresh accident and was laid up once more in his cottage, and from that time until his death he did no more farm work. Joan and her little brother lived or slept at home and worked to keep themselves and him.

Now in this, her own little story, and in her account of the condition of the people at that time; also in the histories of other old men and women whose memories go back as far as hers, supplemented by a little reading in the newspaper of that day, I can understand how it came about that these poor labourers, poor, spiritless slaves as they had been made by long years of extremest poverty and systematic oppression, rose at last against their hard masters and smashed the agricultural machines, and burnt ricks and broke into houses to destroy and plunder their contents. It was a desperate, a mad adventure—these gatherings of half-starved yokels, armed with sticks and axes, and they were quickly put down and punished in a way that even William the Bastard would not have considered as too lenient. But oppression had made them mad; the introduction of thrashing machines was but the last straw, the culminating act of the hideous system followed by landlords and their tenants—the former to get the highest possible rent for his land, the other to get his labour at the lowest possible rate. It was a compact between landlord and tenant aimed against the labourer. It was not merely the fact that the wages of a strong man was only seven shillings a week at the outside, a sum barely sufficient to keep him and his family from starvation and rags (as a fact it was not enough, and but for a little poaching and stealing he could not have lived), but it was customary, especially on the small farms, to get rid of the men after the harvest and leave them to exist the best way they could during the bitter winter months. Thus every village, as a rule, had its dozen or twenty or more men thrown out each year—good steady men, with families dependent on them; and besides these there were the aged and weaklings and the lads who had not yet got a place. The misery of these out-of-work labourers was extreme. They would go to the woods and gather faggots of dead wood, which they would try to sell in the villages; but there were few who could afford to buy of them; and at night they would skulk about the fields to rob a swede or two to satisfy the cravings of hunger.

In some parishes the farmer overseers were allowed to give relief work—out of the rates, it goes without saying—to these unemployed men of the village who had been discharged in October or November and would be wanted again when the winter was over. They would be put to flint-gathering in the fields, their wages being four shillings a week. Some of the very old people of Winterbourne Bishop, when speaking of the principal food of the labourers at that time, the barley bannock and its exceeding toughness, gave me an amusing account of a game of balls invented by the flint-gatherers, just for the sake of a little fun during their long weary day in the fields, especially in cold, frosty weather. The men would take their dinners with them, consisting of a few barley balls or cakes, in their coat pockets, and at noon they would gather at one spot to enjoy their meal, and seat themselves on the ground in a very wide circle, the men about ten yards apart, then each one would produce his bannocks and start throwing, aiming at some other man's face; there were hits and misses and great excitement and hilarity for twenty or thirty minutes, after which the earth and gravel adhering to the balls would be wiped off, and they would set themselves to the hard task of masticating and swallowing the heavy stuff.

At sunset they would go home to a supper of more barley bannocks, washed down with hot water flavoured with some aromatic herb or weed, and then straight to bed to get warm, for there was little firing.

It was not strange that sheep-stealing was one of the commonest offences against the law at that time, in spite of the dreadful penalty. Hunger made the people reckless. My old friend Joan, and other old persons, have said to me that it appeared in those days that the men were strangely indifferent and did not seem to care whether they were hanged or not. It is true they did not hang very many of them—the judge, as a rule, after putting on his black cap and ordering them to the gallows, would send in a recommendation to mercy for most of them; but the mercy of that time was like that of the wicked, exceedingly cruel. Instead of swinging, it was transportation for life, or for fourteen, and at the very least, seven years. Those who have read Clarke's terrible book 'For the Term of His Natural Life' know (in a way) what these poor Wiltshire labourers, who in most cases were never more heard of by their wives and children, were sent to endure in Australia and Tasmania.

And some were hanged; my friend Joan named some people she knows in the neighbourhood who are the grandchildren of a young man with a wife and family of small children who was hanged at Salisbury. She had a vivid recollection of this case because it had seemed so hard, the man having been maddened by want when he took a sheep; also because when he was hanged his poor young wife travelled to the place of slaughter to beg for his body, and had it brought home and buried decently in the village churchyard.

How great the temptation to steal sheep must have been, anyone may know now by merely walking about among the fields in this part of the country to see how the sheep are folded and left by night unguarded, often at long distances from the village, in distant fields and on the downs. Even in the worst times it was never customary, never thought necessary, to guard the flock by night. Many cases could be given to show how easy it was to steal sheep. One quite recent, about twenty years ago, is of a shepherd who was frequently sent with sheep to the fairs, and who on his way to Wilton fair with a flock one night turned aside to open a fold and let out nineteen sheep. On arriving at the fair he took out the stolen sheep and sold them to a butcher of his acquaintance who sent them up to London. But he had taken too many from one flock; they were quickly missed, and by some lucky chance it was found out and the shepherd arrested. He was sentenced to eight months' hard labour, and it came out during the trial that this poor shepherd, whose wages were fourteen shillings a week had a sum of £400 to his credit in a Salisbury bank!

Another case which dates far back is that of a farmer named Day, who employed a shepherd or drover to take sheep to the fairs and markets and steal sheep for him on the way. It is said that he went on at this game for years before it was discovered. Eventually master and man quarrelled and the drover gave information, whereupon Day was arrested and lodged in Fisherton Jail at Salisbury. Later he was sent to take his trial at Devizes, on horseback, accompanied by two constables. At the 'Druid's Head,' a public-house on the way, the three travellers alighted for refreshments, and there Day succeeded in giving them the slip, and jumping on a fast horse, standing ready saddled for him, made his escape. Farmer Day never returned to the Plain and was never heard of again.

There is an element of humour in some of the sheep-stealing stories of the old days. At one village where I stayed often, I heard about a certain Ebenezer Garlick, who was commonly called, in allusion no doubt to his surname, 'Sweet Vi'lets.' He was a sober, hard-working man, an example to most, but there was this against him, that he cherished a very close friendship with a poor, disreputable, drunken loafer nicknamed 'Flitter-mouse,' who spent most of his time hanging about the old coaching inn at the place for the sake of tips. Sweet Vi'lets was always giving coppers and sixpences to this man, but one day they fell out when Flittermouse begged for a shilling. He must, he said, have a

shilling, he couldn't do with less, and when the other refused he followed him, demanding the money with abusive words, to everybody's
astonishment. Finally Sweet Vi'lets turned on him and told him to go
to the devil. Flittermouse in a rage went straight to the constable and
denounced his patron as a sheep-stealer. He, Flittermouse, had been his
servant and helper, and on the very last occasion of stealing a sheep
he had got rid of the skin and offal by throwing them down an old dis-
used well at the top of the village street. To the well the constable
went with ropes and hooks, and succeeded in fishing up the remains
described, and he thereupon arrested Garlick and took him before a
magistrate, who committed him for trial. Flittermouse was the only
witness for the prosecution, and the judge in his summing up said
that, taking into consideration Garlick's known character in the village as a sober, diligent, honest man, it would be a little too much to
hang him on the unsupported testimony of a creature like Flittermouse, who was half fool and half scoundrel. The jury, pleased and
very much surprised at being directed to let a man off, obediently returned a verdict of Not Guilty, and Sweet Vi'lets returned from Salisbury triumphant, to be congratulated on his escape by all the villagers,
who, however, slyly winked and smiled at one another.

Of sheep-stealing stories I will relate one more—a case which never
came into court and was never discovered. It was related to me by a
middle-aged man, a shepherd of Warminster, who had it from his
father, a shepherd of Chitterne, one of the lonely, isolated villages on
Salisbury Plain, between the Avon and the Wylye. His father had it
from the person who committed the crime and was anxious to tell it
to some one, and knew that the shepherd was his true friend, a silent,
safe man. He was a farm-labourer, named Shergold—one of the South
Wiltshire surnames very common in the early part of last century,
which now appear to be dying out—described as a very big, powerful
man, full of life and energy. He had a wife and several young children
to keep, and the time was near mid-winter; Shergold was out of work,
having been discharged from the farm at the end of the harvest; it
was an exceptionally cold season and there was no food and no firing in the house.

One evening in late December a drover arrived at Chitterne with a
flock of sheep which he was driving to Tilshead, another downland
village several miles away. He was anxious to get to Tilshead that
night and wanted a man to help him. Shergold was on the spot and

139

undertook to go with him for the sum of fourpence. They set out when it was getting dark; the sheep were put on the road, the drover going before the flock and Shergold following at the tail. It was a cold, cloudy night, threatening snow, and so dark that he could hardly distinguish the dim forms of even the hindmost sheep, and by and by the temptation to steal one assailed him. For how easy it would be for him to do it! With his tremendous strength he could kill and hide a sheep very quickly without making any sound whatever to alarm the drover. He was very far ahead; Shergold could judge the distance by the sound of his voice when he uttered a call or shout from time to time, and by the barking of the dog, as he flew up and down, first on one side of the road, then on the other, to keep the flock well on it. And he thought of what a sheep would be to him and to his hungry ones at home until the temptation was too strong, and suddenly lifting his big, heavy stick he brought it down with such force on the head of a sheep as to drop it with its skull crushed, dead as a stone. Hastily picking it up he ran a few yards away, and placed it among the furze-bushes, intending to take it home on his way back, and then returned to the flock.

They arrived at Tilshead in the small hours, and after receiving his fourpence he started for home, walking rapidly and then running to be in time, but when he got back to where the sheep was lying the dawn was coming, and he knew that before he could get to Chitterne with that heavy burden on his back people would be getting up in the village and he would perhaps be seen. The only thing to do was to hide the sheep and return for it on the following night. Accordingly he carried it away a couple of hundred yards to a pit or small hollow in the down full of bramble and furze-bushes, and here he concealed it, covering it with a mass of dead bracken and herbage, and left it. That afternoon the long-threatening snow began to fall, and with snow on the ground he dared not go to recover his sheep, since his footprints would betray him; he must wait once more for the snow to melt. But the snow fell all night, and what must his feelings have been when he looked at it still falling in the morning and knew that he could have gone for the sheep with safety, since all traces would have been quickly obliterated!

Once more there was nothing to do but wait patiently for the snow to cease falling and for the thaw. But how intolerable it was; for the weather continued bitterly cold for many days, and the whole country was white. During those hungry days even that poor comfort of sleeping or dozing away the time was denied him, for the danger of discovery was ever present to his mind, and Shergold was not one of the callous men who had become indifferent to their fate; it was his first crime, and he loved his own life and his wife and children, crying to him for food. And the food for them was lying there on the down, close by, and he could not get it! Roast mutton, boiled mutton—mutton in a dozen delicious forms—the thought of it was as distressing, as maddening, as that of the peril he was in.

It was a full fortnight before the wished thaw came; then with fear and trembling he went for his sheep, only to find that it had been pulled to pieces and the flesh devoured by dogs and foxes!

From these memories of the old villagers I turn to the newspapers of the day to make a few citations.

The law as it was did not distinguish between a case of the kind just related, of the starving, sorely tempted Shergold, and that of the systematic thief: sheep-stealing was a capital offence and the man must hang, unless recommended to mercy, and we know what was meant by 'mercy' in those days. That so barbarous a law existed within memory of people to be found living in most villages appears almost incredible to us; but despite the recommendations to 'mercy' usual in a large majority of cases, the law of that time was not more horrible than the temper of the men who administered it. There are good and bad among all, and in all professions, but there is also a black spot in most, possibly in all hearts, which may be developed to almost any extent, and change the justest, wisest, most moral men into 'human devils'—the phrase invented by Canon Wilberforce in another connexion. In reading the old reports and the expressions used by the judges in their summings up and sentences, it is impossible not to believe that the awful power they possessed, and its constant exercise, had not only produced the inevitable hardening effect, but had made them cruel in the true sense of the word. Their pleasure in passing dreadful sentences was very thinly disguised, indeed, by certain lofty conventional phrases as to the necessity of upholding the law, morality, and religion; they were, indeed, as familiar with the name of the Deity as any ranter in a conventicle, and the 'enormity of the crime' was an expression as constantly used in the case of the theft of a loaf of bread, or of an old coat left hanging on a hedge by some ill-clad, half-starved wretch, as in cases of burglary, arson, rape, and murder.

It is surprising to find how very few the real crimes were in those days, despite the misery of the people; that nearly all the 'crimes' for which men were sentenced to the gallows and to transportation for life, or for long terms, were offences which would now be sufficiently punished by a few weeks', or even a few days', imprisonment. Thus in April, 1825, I note that Mr. Justice Park commented on the heavy appearance of the calendar. It was not so much the number (170) of the offenders that excited his concern as it was the nature of the crimes with which they were charged. The worst crime in this instance was sheep-stealing!

Again, this same Mr. Justice Park, at the Spring Assizes at Salisbury, 1827 said that though the calendar was a heavy one, he was happy to find on looking at the depositions of the principal cases, that they were not of a very serious character. Nevertheless he passed sentence of death on twenty-eight perons, among them being one for stealing half a crown!

Of the twenty-eight all but three were eventually reprieved, one of the fated three being a youth of 19, who was charged with stealing a mare and pleaded guilty in spite of a warning from the judge not to do so. This irritated the great man who had the power of life and death in his hand. In passing sentence the judge 'expatiated on the prevalence of the crime of horse-stealing and the necessity of making an example. The enormity of Read's crime rendered him a proper example, and he would therefore hold out no hope of mercy towards him.' As to the plea of guilty, he remarked that nowadays too many persons pleaded guilty, deluded with the hope that it would be taken into consideration and they would escape the severer penalty. He was determined to put a stop to that sort of thing; if Read had not pleaded guilty no doubt some extenuating circumstance would have come up during the trial and he would have saved his life.

There, if ever, spoke the 'human devil' in a black cap!

I find another case of a sentence of transportation for life on a youth of 18, named Edward Baker, for stealing a pocket-handkerchief. Had he pleaded guilty it might have been worse for him.

At the Salisbury Spring Assizes, 1830, Mr. Justice Gazalee, addressing the grand jury, said that none of the crimes appeared to be marked with circumstances of great moral turpitude. The prisoners numbered 130; he passed sentences of death on twenty-nine, life transportations on five, fourteen years on five, seven years on eleven, and various terms of hard labour on the others.

The severity of the magistrates at the quarter-sessions was equally re-volting. I notice in one case, where the leading magistrate on the bench was a great local magnate, an M.P. for Salisbury, etc., a poor fellow with the unfortunate name of Moses Snook was charged with stealing a plank ten feet long, the property of the aforesaid local magnate, M.P., etc., and sentenced to fourteen years' transportation. Sentenced by the man who owned the plank, worth perhaps a shilling or two!

When such was the law of the land and the temper of those who administered it—judges and magistrates or landlords—what must the misery of the people have been to cause them to rise in revolt against their masters! They did nothing outrageous even in the height of their frenzy; they smashed the thrashing machines, burnt some ricks, while the maddest of them broke into a few houses and destroyed their con-tents; but they injured no man; yet they knew what they were fac-ing—the gallows or transportation to the penal settlements ready for their reception at the Antipodes. It is a pity that the history of this ris-ing of the agricultural labourer, the most patient and submissive of men, has never been written. Nothing, in fact, has ever been said of it except from the point of view of landowners and farmers, but there is ample material for a truer and a moving narrative, not only in the brief reports in the papers of the time, but also in the memories of many persons still living, and of their children and children's children, preserved in many a cottage throughout the south of England.

Hopeless as the revolt was and quickly suppressed, it had served to alarm the landlords and their tenants, and taken in conjunction with other outbreaks, notably at Bristol, it produced a sense of anxiety in the mind of the country generally. The feeling found a somewhat amusing expression in the House of Commons, in a motion of Mr. Per-ceval, on 14th February, 1831. This was to move an address to His Majesty to appoint a day for a general fast throughout the United Kingdom. He said that 'the state of the country called for a measure like this—that it was a state of political and religious disorganization—that the elements of the Constitution were being hourly loosened—that in this land there was no attachment, no control, no humility of spirit, no mutual confidence between the poor man and the rich, the employer and the employed; but fear and mistrust and aversion, where, in the time of our fathers, there was nothing but brotherly love and rejoicing before the Lord.'

The House was cynical and smilingly put the matter by, but the anxiety was manifested plainly enough in the treatment meted out to the poor men who had been arrested and were tried before the Special Commissions sent down to Salisbury, Winchester, and other towns. No doubt it was a pleasant time for the judges; at Salisbury thirty-four poor fellows were sentenced to death; thirty-three to be transported for life, ten for fourteen years, and so on.

And here is one last little scene about which the reports in the newspapers of the time say nothing, but which I have from one who witnessed and clearly remembers it, a woman of 95, whose whole life has been passed at a village within sound of the Salisbury Cathedral bells.

It was when the trial was ended, when those who were found guilty and had been sentenced were brought out of the court-house to be taken back to prison, and from all over the Plain and from all parts of Wiltshire their womenfolk had come to learn their fate, and were gathered, a pale, anxious, weeping crowd, outside the gates. The sentenced men came out looking eagerly at the people until they recognized their own and cried out to them to be of good cheer. ' 'Tis hanging for me,' one would say, 'but there'll perhaps be a recommendation to mercy, so don't you fret till you know.' Then another: 'Don't go on so, old mother, 'tis only for life I'm sent.' And yet another: 'Don't you cry, old girl, 'tis only fourteen years I've got, and maybe I'll live to see you all again.' And so on, as they filed out past their weeping women on their way to Fisherton Jail, to be taken thence to the transports in Portsmouth and Plymouth harbours waiting to convey their living freights to that hell on earth so far from home. Not criminals but good, brave men were these!—Wiltshiremen of that strong, enduring, patient class, who not only as labourers on the land but on many a hard-fought field in many parts of the world from of old down to our war of a few years ago in Africa, have shown the stuff that was in them!

But alas! for the poor women who were left—for the old mother who could never hope to see her boy again, and for the wife and her children who waited and hoped against hope through long toiling years.

> And dreamed and started as they slept
> For joy that he was come,

but waking saw his face no more. Very few, so far as I can make out, not more than one in five or six, ever returned.

This, it may be said, was only what they might have expected, the law being what it was—just the ordinary thing. The hideous part of the business was that, as an effect of the alarm created in the minds of those who feared injury to their property and loss of power to oppress the poor labourers, there was money in plenty subscribed to hire witnesses for the prosecution. It was necessary to strike terror into the people. The smell of blood-money brought out a number of scoundrels who for a few pounds were only too ready to swear away the life of any man, and it was notorious that numbers of poor fellows were condemned in this way.

One incident as to this point may be given in conclusion of this chapter about old unhappy things. It relates not to one of those who were sentenced to the gallows or to transportation, but to an inquest and the treatment of the dead.

I have spoken in the last chapter of the mob that visited Hindon, Fonthill, and other villages. They ended their round at Pytt House, near Tisbury, where they broke up the machinery. On that occasion a body of yeomanry came on the scene, but arrived only after the mob had accomplished its purpose of breaking up the thrashing machines. When the troops appeared the 'rioters,' as they were called, made off into the woods and escaped; but before they fled one of them had met his death. A number of persons from the farms and villages around had gathered at the spot and were looking on, when one, a farmer from the neighbouring village of Chilmark, snatched a gun from a gamekeeper's hand and shot one of the rioters, killing him dead. On 27 January, 1831, an inquest was held on the body, and some one was found to swear that the man had been shot by one of the yeomanry, although it was known to everybody that, when the man was shot, the troop had not yet arrived on the scene. The man, this witness stated, had attacked, or threatened, one of the soldiers with his stick, and had been shot. This was sufficient for the coroner; he instructed his jury to bring in a verdict of 'Justifiable homicide,' which they obediently did. 'This verdict,' the coroner then said, 'entailed the same consequences as an act of *felo-de-se*, and he felt that he could not give a warrant for the burial of the deceased. However painful the duty devolved on him in thus adding to the sorrows of the surviving relations, the law appeared too clear to him to admit of an alternative.'

The coroner was just as eager as the judges to exhibit his zeal for the gentry, who were being injured in their interests by these disturbances; and though he could not hang anybody, being only a coroner, he could at any rate kick the one corpse brought before him. Doubtless the 'surviving relations,' for whose sorrows he had expressed sympathy, carried the poor murdered man off by night to hide him somewhere in the earth.

After the law had been thus vindicated and all the business done with, even to the corpse-kicking by the coroner, the farmers were still anxious, and began to show it by holding meetings and discussions on the condition of the labourers. Everybody said that the men had been very properly punished; but at the same time it was admitted that they had some reason for their discontent, that, with bread so dear, it was hardly possible for a man with a family to support himself on seven shillings a week, and it was generally agreed to raise the wages one shilling. But by and by when the anxiety had quite died out, when it was found that the men were more submissive than they had ever been, the lesson they had received having sunk deep into their minds, they cut off the extra shilling, and wages were what they had been—seven shillings a week for a hard-working seasoned labourer, with a family to keep, and from four to six shillings for young unmarried men and for women, even for those who did as much work in the field as any man.

But there were no more risings.

XIX

THE SHEPHERD'S RETURN

*Yarnborough Castle sheep-fair—Caleb leaves Doveton and goes into Dorset
—A land of strange happenings—He is home-sick and returns to Winter-
bourne Bishop—Joseph, his brother, leaves home—His meeting with
Caleb's old master—Settles in Dorset and is joined by his sister Hannah—
They marry and have children—I go to look for them—Joseph Bawcombe
in extreme old age—Hannah in decline*

CALEB'S shepherding period in Doveton came to a somewhat sudden
conclusion. It was nearing the end of August and he was beginning to
think about the sheep which would have to be taken to the 'Castle'
sheep-fair on 5 October, and it appeared strange to him that his master had so far said nothing to him on the subject. By 'Castle' he
meant Yarnborough Castle, the name of a vast prehistoric earth-work
on one of the high downs between Warminster and Amesbury. There
is no village there and no house near; it is nothing but an immense circular wall and trench, inside of which the fair is held. It was formerly
one of the most important sheep-fairs in the country, but for the last
two or three decades has been falling off and is now of little account.
When Bawcombe was shepherd at Doveton it was still great, and
when he first went there as Mr. Ellerby's head-shepherd he found himself regarded as a person of considerable importance at the Castle.
Before setting out with the sheep he asked for his master's instruc-
tions and was told that when he got to the ground he would be directed by the persons in charge to the proper place. The Ellerbys, he
said, had exhibited and sold their sheep there for a period of eighty-
eight years, without missing a year, and always at the same spot.
Every person visiting the fair on business knew just where to find the
Ellerbys' sheep, and he added with pride, they expected them to be
the best sheep at the Castle.

One day Mr. Ellerby came to have a talk with his shepherd, and in
reply to a remark of the latter about the October sheep-fair he said that
he would have no sheep to send. 'No sheep to send, master!' exclaimed
Caleb in amazement. Then Mr. Ellerby told him that he had taken a
notion into his head that he wanted to go abroad with his wife for a

time, and that some person had just made him so good an offer for all his sheep that he was going to accept it, so that for the first time in eighty-eight years there would be no sheep from Doveton Farm at the Castle fair. When he came back he would buy again; but if he could live away from the farm, he would probably never come back—he would sell it.

Caleb went home with a heavy heart and told his wife. It grieved her, too, because of her feeling for Mrs. Ellerby, but in a little while she set herself to comfort him. 'Why, what's wrong about it?' she asked. ' 'Twill be more'n three months before the year's out, and master'll pay for all the time sure, and we can go home to Bishop and bide a little without work, and see if that father of yours has forgiven 'ee for going away to Warminster.'

So they comforted themselves, and were beginning to think with pleasure of home when Mr. Ellerby informed his shepherd that a friend of his, a good man though not a rich one, was anxious to take him as head-shepherd, with good wages and a good cottage rent free. The only drawback for the Bawcombes was that it would take them still farther from home, for the farm was in Dorset, although quite near the Wiltshire border.

Eventually they accepted the offer, and by the middle of September were once more settled down in what was to them a strange land. How strange it must have seemed to Caleb, how far removed from home and all familiar things, when even to this day, more than forty years later, he speaks of it as the ordinary modern man might speak of a year's residence in Uganda, Tierra del Fuego, or the Andaman Islands! It was a foreign country, and the ways of the people were strange to him, and it was a land of very strange things. One of the strangest was an old ruined church in the neighbourhood of the farm where he was shepherd. It was roofless, more than half fallen down, and all the standing portion, with the tower, overgrown with old ivy; the building itself stood in the centre of a huge round earthwork and trench, with large barrows on the ground outside the circle. Concerning this church he had a wonderful story: its decay and ruin had come about after the great bell in the tower had mysteriously disappeared, stolen one stormy night, it was believed, by the Devil himself. The stolen bell, it was discovered, had been flung into a small river at a distance of some miles from the church, and there in summer-time, when the water was low, it could be distinctly seen lying half buried in the mud at the bottom. But all the king's horses and all the king's men couldn't pull it out; the Devil, who pulled the other way, was strongest. Eventually some wise person said that a team of white oxen would be able to pull it out, and after much seeking the white oxen were obtained, and thick ropes were tied to the sunken bell, and the cattle were goaded and yelled at, and tugged and strained until the bell came up and was finally drawn right up to the top of the steep, cliff-like bank of the stream. Then one of the teamsters shouted in triumph, 'Now we've got out the bell, in spite of all the devils in hell,' and no sooner had he spoken the bold words than the ropes parted, and back tumbled the bell to its old place at the bottom of the river, where it remains to this day. Caleb had once met a man in those parts who assured him that he had seen the bell with his own eyes, lying nearly buried in mud at the bottom of the stream.

The legend is not in the history of Dorset; a much more prosaic account of the disappearance of the bell is there given, in which the Devil took no part unless he was at the back of the bad men who were concerned in the business. But in this strange, remote country, outside of 'Wiltsheer,' Bawcombe was in a region where anything might have happened, where the very soil and pasture was unlike that of his native country, and the mud adhered to his boots in a most unaccountable way. It was almost uncanny. Doubtless he was home-sick, for a month or two before the end of the year he asked his master to look out for another shepherd.

This was a great disappointment to the farmer: he had gone a distance from home to secure a good shepherd, and had hoped to keep him permanently, and now after a single year he was going to lose him. What did the shepherd want? He would do anything to please him, and begged him to stay another year. But no, his mind was set on going back to his own native village and to his own people. And so when his long year was ended he took his crook and set out over the hills and valleys, followed by a cart containing his 'sticks' and wife and children. And at home with his old parents and his people he was happy once more; in a short time he found a place as head-shepherd, with a cottage in the village, and followed his flock on the old familiar down, and everything again was as it had been from the beginning of life and as he desired it to be even to the end.

His return resulted incidentally in other changes and migrations in the Bawcombe family. His elder brother Joseph, unmarried still although his senior by about eight years, had not got on well at home. He was a person of a peculiar disposition, so silent with so fixed and unsmiling an expression, that he gave the idea of a stolid, thick-skinned man, but at bottom he was of a sensitive nature, and feeling that his master did not treat him properly, he gave up his place and was for a long time without one. He was singularly attentive to all that fell from Caleb about his wide wanderings and strange exper-iences, especially in the distant Dorset country; and at length, about a year after his brother's return, he announced his intention of going away from his native place for good to seek his fortune in some distant place where his services would perhaps be better appreciated. When asked where he intended going, he answered that he was going to look for a place in that part of Dorset where Caleb had been shepherd for a year and had been so highly thought of.

Now Joseph, being a single man, had no 'sticks'; all his possessions went into a bundle, which he carried tied to his crook, and with his sheep-dog following at his heels he set forth early one morning on the most important adventure of his life. Then occurred an instance of what we call a coincidence, but which the shepherd of the downs, nursed in the old beliefs and traditions, prefers to regard as an act of providence.

About noon he was trudging along in the turnpike road when he was met by a farmer driving in a trap, who pulled up to speak to him and asked him if he could say how far it was to Winterbourne Bishop. Joseph replied that it was about fourteen miles—he had left Bishop that morning.

Then the farmer asked him if he knew a man there named Caleb Bawcombe, and if he had a place as shepherd there, as he was now on his way to look for him and to try and persuade him to go back to Dorset, where he had been his head-shepherd for the space of a year.

Joseph said that Caleb had a place as head-shepherd on a farm at Bishop, that he was satisfied with it, and was moreover one that pre-ferred to bide in his native place.

The farmer was disappointed, and the other added, 'Maybe you've heard Caleb speak of his elder brother Joseph—I be he.'

'What!' exclaimed the farmer. 'You're Caleb's brother! Where be going then?—to a new place?

'I've got no place; I be going to look for a place in Dorsetsheer.'

'Tis strange to hear you say that,' exclaimed the farmer. He was

going, he said, to see Caleb, and if he would not or could not go back
to Dorset himself to ask him to recommend some man of the village to
him; for he was tired of the ways of the shepherds of his own part of
the country, and his heart was set on getting a man from Caleb's village, where shepherds understood sheep and knew their work. 'Now
look here, shepherd,' he continued, 'if you'll engage yourself to me for
a year I'll go no farther, but take you right back with me in the trap.'

The shepherd was very glad to accept the offer; he devoutly believed that in making it the farmer was but acting in accordance with
the will of a Power that was mindful of man and kept watch on him,
even on His poor servant Joseph, who had left his home and people
to be a stranger in a strange land.

So well did servant and master agree that Joseph never had occasion
to look for another place; when his master died an old man, his son
succeeded him as tenant of the farm, and he continued with the son
until he was past work. Before his first year was out, his younger sister, Hannah, came to live with him and keep house, and eventually
they both got married, Joseph to a young woman of the place, and
Hannah to a small working farmer whose farm was about a mile from
the village. Children were born to both, and in time grew up, Joseph's
sons following their father's vocation, while Hannah's were brought
up to work on the farm. And some of them, too, got married in time
and had children of their own.

These are the main incidents in the lives of Joseph and Hannah, related to me at different times by their brother; he had followed their
fortunes from a distance, sometimes getting a message, or hearing of
them incidentally, but he did not see them. Joseph never returned to
his native village, and the visits of Hannah to her old home had been
few and had long ceased. But he cherished a deep enduring affection
for both; he was always anxiously waiting and hoping for tidings of
them, for Joseph was now a feeble old man living with one of his
sons, and Hannah, long a widow, was in declining health, but still kept
the farm, assisted by one of her sons and two unmarried daughters.
Though he had not heard for a long time it never occurred to him to
write, nor did they ever write to him.

Then, when I was staying at Winterbourne Bishop and had the intention of shortly paying a visit to Caleb, it occurred to me one day to
go into Dorset and look for these absent ones, so as to be able to give
him an account of their state. It was a long journey, and arrived

at the village I soon found a son of Joseph, a fine-looking man, who took me to his cottage, where his wife led me into the old shepherd's room. I found him very aged in appearance, with a grey face and sunken cheeks, lying on his bed and breathing with difficulty; but when I spoke to him of Caleb a light of joy came into his eyes, and he raised himself on his pillows, and questioned me eagerly about his brother's state and family, and begged me to assure Caleb that he was still quite well, although too feeble to get about much, and that his children were taking good care of him.

From the old brother I went on to seek the young sister—there was a difference of more than twenty years in their respective ages—and found her at dinner in the large old farm-house kitchen. At all events she was presiding, the others present being her son, their hired labourer, the farm boy, and two unmarried daughters. She herself tasted no food. I joined them at their meal, and it gladdened and saddened me at the same time to be with this woman, for she was Caleb's sister, and was attractive in herself, looking strangely young for her age, with beautiful dark, soft eyes and but few white threads in her abundant black hair. The attraction was also in her voice and speech and manner; but alas! there was that in her face which was painful to witness—the signs of long suffering, of nights that bring no refreshment, an expression in the eyes of one that is looking anxiously out into the dim distance—a vast unbounded prospect, but with clouds and darkness resting on it.

It was not without a feeling of heaviness at the heart that I said good-bye to her, nor was I surprised when, less than a year later, Caleb received news of her death.

XX

THE DARK PEOPLE OF THE VILLAGE

How the materials for this book were obtained—The hedgehog-hunter—A gipsy taste—History of a dark-skinned family—Hedgehog eaters—Half-bred and true gipsies—Perfect health—Eating carrion—Mysterious knowl-edge and faculties—The three dark Wiltshire types—Story of another dark man of the village—Account of Liddy—His shepherding—A happy life with horses—Dies of a broken heart—His daughter

I HAVE sometimes laughed to myself when thinking how a large part of the material composing this book was collected. It came to me in conversations, at intervals, during several years, with the shepherd. In his long life in his native village, a good deal of it spent on the quiet down, he had seen many things it was or would be interesting to hear; the things which had interested him, too, at the time, and had fallen into oblivion, yet might be recovered. I discovered that it was of little use to question him: the one valuable recollection he possessed on any subject would, as a rule, not be available when wanted; it would lie just beneath the surface so to speak, and he would pass and repass over the ground without seeing it. He would not know that it was there; it would be like the acorn which a jay or squirrel has hidden and forgotten all about, which he will nevertheless recover some day if by chance something occurs to remind him of it. The only method was to talk about the things he knew, and when by chance he was reminded of some old experience or some little observation or incident worth hearing, to make a note of it, then wait patiently for something else. It was a very slow process, but it is not unlike the one we practise always with regard to wild nature. We are not in a hurry, but are always watchful, with eyes and ears and mind open to what may come; it is a mental habit, and when nothing comes we are not disappointed—the act of watching has been a sufficient pleasure; and when something does come we take it joyfully as if it were a gift—a valuable object picked up by chance in our walks.

When I turned into the shepherd's cottage, if it was in winter and he was sitting by the fire, I would sit and smoke with him, and if we were in a talking mood I would tell him where I had been and what I

had heard and seen, on the heath, in the woods, in the village, or anywhere, on the chance of its reminding him of something worth hearing in his past life.

One Sunday morning, in the late summer, during one of my visits to
him, I was out walking in the woods and found a man of the village, a
farm labourer, with his small boy hunting for hedgehogs. He had
caught and killed two which the boy was carrying. He told me he
was very found of the flesh of hedgehogs—'pigs', he called them for
short; he said he would not exchange one for a rabbit. He always
spent his holidays pig-hunting; he had no dog and didn't want one; he
found them himself, and his method was to look for the kind of place
in which they were accustomed to live—a thick mass of bramble
growing at the side of an old ditch as a rule. He would force his way
into it and, moving round and round, trample down the roots and
loose earth and dead leaves with his heavy iron-shod boots until he
broke into the nest or cell of the spiny little beast hidden away under
the bush.

He was a short, broad-faced man, with a brown skin, black hair,
and intensely black eyes. Talking with the shepherd that evening I
told him of the encounter, and remarked that the man was probably a
gipsy in blood, although a labourer, living in the village and married
to a woman with blue eyes who belonged to the place.

This incident reminded him of a family, named Targett, in his
native village, consisting of four brothers and a sister. He knew them
first when he was a boy himself, but could not remember their parents. 'It seemed as if they didn't have any,' he said. The four brothers
were very much alike: short, with broad faces, black eyes and hair,
and brown skins. They were good workers, but somehow they were
never treated by the farmers like the other men. They were paid
less wages—as much as two to four shillings a week less per man—
and made to do things that others would not do, and generally imposed upon. It was known to every employer of labour in the place
that they could be imposed upon; yet they were not fools, and occasionally if their master went too far in bullying and abusing them
and compelling them to work overtime every day, they would have
sudden violent outbursts of rage and go off without any pay at all.
What became of their sister he never knew: but none of the four
brothers ever married; they lived together always, and two died in the
village, the other two going to finish their lives in the workhouse.

One of the curious things about these brothers was that they had a passion for eating hedgehogs. They had it from boyhood, and as boys used to go a distance from home and spend the day hunting in hedges and thickets. When they captured a hedgehog they would make a small fire in some sheltered spot and roast it, and while it was roasting one of them would go to the nearest cottage to beg for a pinch of salt, which was generally given.

These, too, I said, must have been gipsies, at all events on one side. Where there is a cross the gipsy strain is generally strongest, although the children, if brought up in the community, often remain in it all their lives; but they are never quite of it. Their love of wildness and of eating wild flesh remains in them, and it is also probable that there is an instability of character, a restlessness, which the small farmers who usually employ such men know and trade on; the gipsy who takes to farm work must not look for the same treatment as the big-framed, white-skinned man who is as strong, enduring, and unchangeable as a draught horse or ox, and constant as the sun itself.

The gipsy element is found in many if not most villages in the south of England. I know one large scattered village where it appears predominant—as dirty and disorderly-looking a place as can be imagined, the ground round every cottage resembling a gipsy camp, but worse owing to its great litter of old rags and rubbish strewn about. But the people, like all gipsies, are not so poor as they look, and most of the cottagers keep a trap and pony with which they scour the country for many miles around in quest of bones, rags, and bottles, and anything else they can buy for a few pence, also anything they can 'pick up' for nothing.

This is almost the only kind of settled life which a man with a good deal of gipsy blood in him can tolerate; it affords some scope for his chaffering and predatory instincts and satisfies the roving passion, which is not so strong in those of mixed blood. But it is too respectable or humdrum a life for the true, undegenerate gipsy. One wet evening in September last I was prowling in a copse near Shrewton, watching the birds, when I encountered a young gipsy and recognized him as one of a gang of about a dozen I had met several days before near Salisbury. They were on their way, they had told me, to a village near Shaftesbury, where they hoped to remain a week or so.

'What are you doing here?' I asked my gipsy.

He said he had been to Idmiston; he had been on his legs out in the rain and wet to the skin since morning. He didn't mind that much as the wet didn't hurt him and he was not tired; but he had eight miles to walk yet over the downs to a village on the Wylye where his people were staying.

I remarked that I had thought they were staying over Shaftesbury way.

He then looked sharply at me. 'Ah, yes', he said, 'I remember we met you and had some talk a fortnight ago. Yes, we went there, but they wouldn't have us. They soon ordered us off. They advised us to settle down if we wanted to stay anywhere. Settle down! I'd rather be dead!'

There spoke the true gipsy; and they are mostly of that mind. But what a mind it is for human beings in this climate! It is in a year like this of 1909, when a long cold winter and a miserable spring, with frosty nights lasting well into June, was followed by a cold wet summer and a wet autumn, that we can see properly what a mind and body is his—how infinitely more perfect the correspondence between organism and environment in his case than in ours, who have made our own conditions, who have not only houses to live in, but a vast army of sanitary inspectors, physicians and bacteriologists to safeguard us from that wicked stepmother who is anxious to get rid of us before our time! In all this miserable year, during which I have met and conversed with and visited many scores of gipsies, I have not found one who was not in a cheerful frame of mind, even when he was under a cloud with the police on his track; nor one with a cold, or complaining of an ache in his bones, or of indigestion.

The subject of gipsies catching cold connects itself just now in my mind with that of the gipsy's sense of humour. He has that sense, and it makes him happy when he is reposing in the bosom of his family and can give it free vent; but the instant you appear on the scene its gracious outward signs vanish like lightning and he is once more the sly, subtle animal, watching you furtively, but with the intensity of Gip the cat, described in a former chapter. When you have left him and he relaxes the humour will come back to him; for it is a humour similar to that of some of the lower animals, especially birds of the crow family, and of primitive people, only more highly developed, and is concerned mainly with the delight of trickery—with getting the better of some one and the huge enjoyment resulting from the process.

One morning, between nine and ten o'clock, during the excessively cold spell near the end of November, 1909, I paid a visit to some gipsies I knew at their camp. The men had already gone off for the day, but some of the women were there—a young married woman, two big girls, and six or seven children. It was a hard frost and their sleeping accommodation was just as in the summer-time—bundles of straw and old rugs placed in or against little half-open canvas and rag shelters; but they all appeared remarkably well, and some of the children were standing on the hard frozen ground with bare feet. They assured me that they were all well, that they hadn't caught colds and didn't mind the cold. I remarked that I had thought the severe frost might have proved too much for some of them in that high, unsheltered spot in the downs, and that if I had found one of the children down with a cold I should have given it a sixpence to comfort it. 'Oh,' cried the young married woman, 'there's my poor six months' old baby half dead of a cold; he's very bad, poor dear, and I'm in great trouble about him.'

'He is bad, the darling!' cried one of the big girls. 'I'll soon show you how bad he is!' and with that she dived into a pile of straw and dragged out a huge fat sleeping baby. Holding it up in her arms she begged me to look at it to see how bad it was; the fat baby slowly opened its drowsy eyes and blinked at the sun, but uttered no sound, for it was not a crying baby, but was like a great fat retriever pup pulled out of its warm bed.

How healthy they are is hardly known even to those who make a special study of these aliens, who, albeit aliens, are yet more native than any Englishman in the land. It is not merely their indifference to wet and cold; more wonderful still is their dog-like capacity of assimilating food which to us would be deadly. This is indeed not a nice or pretty subject, and I will give but one instance to illustrate my point; the reader with a squeamish stomach may skip the ensuing paragraph.

An old shepherd of Chitterne relates that a family, or gang, of gipsies used to turn up from time to time at the village; he generally saw them at lambing-time, when one of the heads of the party with whom he was friendly would come round to see what he had to give them. On one occasion his gipsy friend appeared, and after some conversation on general subjects, asked him if he had anything in his way. 'No, nothing this time,' said the shepherd. 'Lambing was over two or three months ago and there's nothing left—no dead lamb. I hung up a few cauls on a beam in the old shed, thinking they would do for the dogs, but forgot them and they went bad and then dried.'

'They'll do very well for us,' said his friend.

'No, don't you take them!' cried the shepherd in alarm; 'I tell you they went bad months ago, and 'twould kill anyone to eat such stuff. They've dried up now, and are dry and black as old skin.'

'That doesn't matter—we know how to make them all right,' said the gipsy. 'Soaked with a little salt, then boiled, they'll do very well.' And off he carried them.

In reading the reports of the Assizes held at Salisbury from the late eighteenth century down to about 1840, it surprised me to find how rarely a gipsy appeared in that long, sad, monotonous procession of 'criminals' who passed before the man sitting with his black cap on his head, and were sent to the gallows or to the penal settlements for stealing sheep and fowls and ducks or anything else. Yet the gipsies were abundant then as now, living the same wild, lawless life, quartering the country, and hanging round the villages to spy out everything stealable. The man caught was almost invariably the poor, slow-minded, heavy-footed agricultural labourer; the light, quick-moving, cunning gipsy escaped. In the 'Salisbury Journal' for 1820 I find a communication on this subject, in which the writer says that a common trick of the gipsies was to dig a deep pit at their camp in which to bury a stolen sheep, and on this spot they would make their camp fire. If the sheep was not missed, or if no report of its loss was made to the police, the thieves would soon be able to dig it up and enjoy it; but if inquiries were made they would have to wait until the affair had blown over.

It amused me to find, from an incident related to me by a workman in a village where I was staying lately, that this simple, ancient device is still practised by the gipsies. My informant said that on going out at about four o'clock one morning during the late summer he was surprised at seeing two gipsies with a pony and cart at the spot where a party of them had been encamped a fortnight before. He watched them, himself unseen, and saw that they were digging a pit on the spot where they had had their fire. They took out several objects from the ground, but he was too far away to make out what they were. They put them in the cart and covered them over, then filled up the pit, trampled the earth well down, and put the ashes and burnt sticks back in the same place, after which they got into the cart and drove off.

Of course a man, even a nomad, must have some place to conceal his treasures or belongings in, and the gipsy has no cellar nor attic nor secret cupboard, and as for his van it is about the last place in which he would bestow anything of value or incriminating, for though he is always on the move, he is, moving or sitting still, always under a cloud. The ground is therefore the safest place to hide things in, especially in a country like the Wiltshire Downs, though he may use rocks and hollow trees in other districts. His habit is that of the jay and magpie, and of the dog with a bone to put by till it is wanted. Possibly the rural police have not yet discovered this habit of the gipsy. Indeed, the contrast in mind and locomotive powers between the gipsy and the village policeman has often amused me; the former most like the thievish jay, ever on mischief bent; the other, who has his eye on him, is more like the portly Cochin-China fowl of the farmyard, or the Muscovy duck, or stately gobbler.

To go back. When the buried sheep had to be kept too long buried and was found 'gone bad' when disinterred, I fancy it made little difference to the diners. One remembers Thoreau's pleasure at the spectacle of a crowd of vultures feasting on the carrion of a dead horse; the fine healthy appetite and boundless vigour of nature filled him with delight. But it is not only some of the lower animals—dogs and vultures, for instance—which possess this power and immunity from the effects of poisons developed in putrid meat; the Greenlanders and African savages, and many other peoples in various parts of the world, have it as well.

Sometimes when sitting with gipsies at their wild hearth, I have felt curious as to the contents of that black pot simmering over the fire. No doubt it often contains strange meats, but it would not have been etiquette to speak of such a matter. It is like the pot on the fire of the Venezuela savage into which he throws whatever he kills with his little poisoned arrows or fishes out of the river. Probably my only quarrel with them would be about the little fledgelings: it angers me to see them beating the bushes in spring in search of small nesties and the callow young that are in them. After all, the gipsies could retort that my friends the jays and magpies are at the same business in April and May.

It is just these habits of the gipsy which I have described, shocking to the moralist and sanitarian and disgusting to the person of delicate stomach, it may be, which please me, rather than the romance and poetry which the scholar-gipsy enthusiasts are fond of reading into him. He is to me a wild, untameable animal of curious habits, and interests me as a naturalist accordingly. It may be objected that being a naturalist occupied with the appearance of things, I must inevitably miss the one thing which others find.

In a talk I had with a gipsy a short time ago, he said to me: 'You know what the books say, and we don't. But we know other things that are not in the books, and that's what we have. It's ours, our own, and you can't know it.'

It was well put; but I was not perhaps so entirely ignorant as he imagined of the nature of that special knowledge, or shall we say faculty, which he claimed. I take it to be cunning—the cunning of a wild animal with a man's brain—and a small, infinitesimal, dose of something else which eludes us. But that something else is not of a spiritual nature: the gipsy has no such thing in him; the soul growths are rooted in the social instinct, and are developed in those in whom that instinct is strong. I think that if we analyse that dose of something else, we will find that it is still the animal's cunning, a special, a sublimated cunning, the fine flower of his whole nature, and that it has nothing mysterious in it. He is a parasite, but free and as well able to exist free as the fox or jackal; but the parasitism pays him well, and he has followed it so long in his intercourse with social man that it has come to be like an instinct, or secret knowledge, and is nothing more than a marvellously keen penetration which reveals to him the character and degree of credulity and other mental weaknesses of his subject.

It is not so much the wind on the heath, brother, as the fascination of lawlessness, which makes his life an everlasting joy to him; to pit himself against game-keeper, farmer, policeman, and everybody else, and defeat them all, to flourish like the parasitic fly on the honey in the hive and escape the wrath of the bees.

I must now return from this long digression to my conversation with the shepherd about the dark people of the village.

There were, I continued, other black-eyed and black-haired people in the villages who had no gipsy blood in their veins. So far as I could make out there were dark people of three originally distinct and widely different races in the Wiltshire Downs. There was a good deal of mixed blood, no doubt, and many dark persons could not be identified as belonging to any particular race. Nevertheless three distinct types could be traced among the dark people, and I took them to be, first, the gipsy, rather short of stature, brown-skinned, with broad face and high cheek-bones, like the men we had just been speaking of. Secondly the men and women of white skins and good features, who had rather broad faces and round heads, and were physically and mentally just as good as the best blue-eyed people; these were probably the descendants of the dark, broad-faced Wilsetæ, who came over at the time when the country was being overrun with the English and other nations or tribes, and who colonized in Wiltshire and gave it their name. The third type differed widely from both the others. They were smallest in size and had narrow heads and long or oval faces, and were very dark, with brown skins; they also differed mentally from the others, being of a more lively disposition and hotter temper. The characters which distinguish the ancient British or Iberian race appeared to predominate in persons of this type.

The shepherd said he didn't know much about 'all that,' but he remembered that they once had a man in the village who was like the last kind I had described. He was a labourer named Tark, who had several sons, and when they were grown up there was a last one born: he had to be the last because his mother died when she gave him birth; and that last one was like his father, small, very dark-skinned, with eyes like sloes, and exceedingly lively and active.

Tark, himself, he said, was the liveliest, most amusing man he had ever known, and the quickest to do things, whatever it was he was asked to do, but he was not industrious and not thrifty. The Tarks were always very poor. He had a good ear for music and was a singer of the old songs—he seemed to know them all. One of his performances was with a pair of cymbals which he had made for himself out of some old metal plates, and with these he used to play while dancing about, clashing them in time, striking them on his head, his breast, and legs. In these dances with the cymbals he would whirl and leap about in an astonishing way, standing sometimes on his hands, then on his feet, so that half the people in the village used to gather at his cottage to watch his antics on a summer evening.

One afternoon he was coming down the village street and saw the blacksmith standing near his cottage looking up at a tall fir-tree which grew there on his ground. 'What be looking at?' cried Tark. The blacksmith pointed to a branch, the lowest branch of all, but about forty feet from the ground, and said a chaffinch had his nest in it, about three feet from the trunk, which his little son had set his heart on having. He had promised to get it down for him, but there was no long ladder and he didn't know how to get it.

Tark laughed and said that for half a gallon of beer he would go up legs first and take the nest and bring it down in one hand, which he would not use in climbing, and would come down as he went up, head first.

'Do it then,' said the blacksmith, 'and I'll stand the half gallon.'

Tark ran to the tree, and turning over and standing on his hands, clasped the bole with his legs and then with his arms and went up to the branch, when taking the nest and holding it in one hand, he came down head first to the ground in safety.

There were other anecdotes of his liveliness and agility. Then followed the story of the youngest son, known as Liddy. 'I don't rightly know,' said Caleb, 'what the name was he was given when they christened'n; but he were always called Liddy, and nobody knowed any other name for him.'

Liddy's grown-up brothers all left home when he was a small boy: one enlisted and was sent to India and never returned; the other two went to America, so it was said. He was twelve years old when his father died, and he had to shift for himself; but he was no worse off on that account, as they had always been very poor owing to poor Tark's love of beer. Before long he got employed by a small working farmer who kept a few cows and a pair of horses and used to buy wethers to fatten them, and these the boy kept on the down.

Liddy was always a 'leetel chap,' and looked no more than 9 when

12, so that he could do no heavy work; but he was a very willing and active little fellow, with a sweet temper, and so lively and full of fun as to be a favourite with everybody in the village. The men would laugh at his pranks, especially when he came from the fields on the old plough horse and urged him to a gallop, sitting with his face to the tail; and they would say that he was like his father, and would never be much good except to make people laugh. But the women had a tender feeling for him, because, although motherless and very poor, he yet contrived to be always clean and neat. He took the greatest care of his poor clothes, washing and mending them himself. He also took an intense interest in his wethers, and almost every day he would go to Caleb, tending his flock on the down, to sit by him and ask a hundred questions about sheep and their management. He looked on Caleb, as head-shepherd on a good-sized farm, as the most important and most fortunate person he knew, and was very proud to have him as guide, philosopher and friend.

Now it came to pass that once in a small lot of thirty or forty wethers which the farmer had bought at a sheep-fair and brought home it was discovered that one was a ewe—a ewe that would perhaps at some future day have a lamb! Liddy was greatly excited at the discovery; he went to Caleb and told him about it, almost crying at the thought that his master would get rid of it. For what use would it be to him? but what a loss it would be! And at last, plucking up courage he went to the farmer and begged and prayed to be allowed to keep the ewe, and the farmer laughed at him; but he was a little touched at the boy's feeling, and at last consented. Then Liddy was the happiest boy in the village, and whenever he got the chance he would go out to Caleb on the down to talk about and give him news of the one beloved ewe. And one day, after about nineteen or twenty weeks, Caleb, out with his flock, heard shouts at a distance, and, turning to look, saw Liddy coming at great speed towards him, shouting out some great news as he ran; but what it was Caleb could not make out, even when the little fellow had come to him, for his excitement made him incoherent. The ewe had lambed, and there were twins—two strong healthy lambs, most beautiful to see! Nothing so wonderful had ever happened in his life before! And now he sought out his friend oftener than ever, to talk of his beloved lambs, and to receive the most minute directions about their care. Caleb, who is not a laughing man, could not help laughing a little when he recalled poor Liddy's enthusiasm. But that beautiful shining chapter in the poor boy's life could not last, and when the lambs were grown they were sold, and so were all the wethers, then Liddy, not being wanted, had to find something else to do.

I was too much interested in this story to let the subject drop. What had been Liddy's after-life? Very uneventful: there was, in fact, nothing in it, nor in him, except an intense love for all things, especially animals; and nothing happened to him until the end, for he has been dead now these nine or ten years. In his next place he was engaged, first, as carter's boy, and then under-carter, and all his love was lavished on the horses. They were more to him than sheep, and he could love them without pain, since they were not being prepared for the butcher with his abhorred knife. Liddy's love and knowledge of horses became known outside of his own little circle, and he was offered and joyfully accepted a place in the stables of a wealthy young gentleman farmer, who kept a large establishment and was a hunting man. From stable-boy he was eventually promoted to groom. Occasionally he would reappear in his native place. His home was but a few miles away, and when out exercising a horse he appeared to find it a pleasure to trot down the old street, where as a farmer's boy he used to make the village laugh at his antics. But he was very much changed from the poor boy, who was often hatless and barefooted, to the groom in his neat, well-fitting black suit, mounted on a showy horse.

In this place he continued about thirty years, and was married and had several children and was very happy, and then came a great disaster. His employer having met with heavy losses sold all his horses and got rid of his servants, and Liddy had to go. This great change, and above all his grief at the loss of his beloved horses, was more than he could endure. He became melancholy and spent his days in silent brooding, and by and by, to everybody's surprise, Liddy fell ill, for he was in the prime of life and had always been singularly healthy. Then to astonish people still more, he died. What ailed him—what killed him? every one asked of the doctor; and his answer was that he had no disease—that nothing ailed him except a broken heart; and that was what killed poor Liddy.

In conclusion I will relate a little incident which occurred several months later, when I was again on a visit to my old friend the shepherd. We were sitting together on a Sunday evening, when his old wife looked out and said, 'Lor, here be Mrs. Taylor with her children coming in to see us.' And Mrs. Taylor soon appeared, wheeling her baby in a perambulator, with two little girls following. She was a comely, round, rosy little woman, with black hair, black eyes, and a singularly sweet expression, and her three pretty little children were like her. She stayed half an hour in pleasant chat, then went her way down the road to her home. Who, I asked, was Mrs. Taylor?

Bawcombe said that in a way she was a native of their old village of Winterbourne Bishop: at least her father was. She had married a man who had taken a farm near them, and after having known her as a young girl they had been glad to have her again as a neighbour. 'She's a daughter of that Liddy I told 'ee about some time ago,' he said.

XXI

SOME SHEEP-DOGS

Breaking a sheep-dog—The shepherd buys a pup—His training—He refuses to work—He chases a swallow and is put to death—The shepherd's remorse—Bob, the sheep-dog—How he was bitten by an adder—Period of the dog's receptivity—Tramp, the sheep-dog—Roaming lost about the country—A rage of hunger—Sheep-killing dogs—Dogs running wild— Anecdotes—A Russian sheep-dog—Caleb parts with Tramp

To CALEB the proper training of a dog was a matter of the very first importance. A man, he considered, must have not only a fair amount of intelligence, but also experience, and an even temper, and a little sympathy a swell, to sum up the animal in hand—its special aptitudes, its limitations, its disposition, and that something in addition, which he called a 'kink,' and would probably have described as its idiosyncrasy if he had known the word. There was as much individual difference among dogs as there is in boys; but if the breed was right, and you went the right way about it, you could hardly fail to get a good servant. If a dog was not properly broken, if its trainer had not made the most of it, he was not a 'good shepherd': he lacked the intelligence—'understanding' was his word—or else the knowledge or patience or persistence to do his part. It was, however, possible for the best shepherd to make mistakes, and one of the greatest to be made, which was not uncommon, was to embark on the long and laborious business of training an animal of mixed blood—a sheep-dog with a taint of terrier, retriever, or some other unsuitable breed in him. In discussing this subject with other shepherds I generally found that those who were in perfect agreement with Caleb on this point were men who were somewhat like him in character, and who regarded their work with the sheep as so important that it must be done thoroughly in every detail and in the best way. One of the best shepherds I know, who is 60 years old and has been on the same downland sheep-farm all his life, assures me that he has never had and never would have a dog which was trained by another. But the shepherd of the ordinary kind says that he doesn't care much about the animal's parentage, or that he doesn't trouble to inquire into its pedigree: he breaks the animal, and finds that he does pretty well, even when he has some strange blood in him; finally, that all dogs have faults and you must put up with them. Caleb would say of such a man that he was not a 'good shepherd.' One of his saddest memories was of a dog which he bought and broke without having made the necessary inquiries about its parentage.

It happened that a shepherd of the village, who had taken a place at a distant farm, was anxious to dispose of a litter of pups before leaving, and he asked Caleb to have one. Caleb refused. 'My dog's old, I know,' he said, 'but I don't want a pup now and I won't have 'n.'

A day or two later the man came back and said he had kept one of the best of the five for him—he had got rid of all the others. 'You can't do better,' he persisted. 'No' said Caleb, 'what I said I say again. I won't have 'n, I've no money to buy a dog.'

'Never mind about money,' said the other. 'You've got a bell I like the sound of; give he to me and take the pup.' And so the exchange was made, a copper bell for a nice black pup with a white collar; its mother, Bawcombe knew, was a good sheep-dog, but about the other parent he made no inquiries.

On receiving the pup he was told that its name was Tory, and he did not change it. It was always difficult, he explained, to find a name for a dog—a name, that is to say, which anyone would say was a proper name for a dog and not a foolish name. One could think of a good many proper names—Jack and Watch, and so on—but in each case one would remember some dog which had been called by that name, and it seemed to belong to that particular well-remembered dog and to no other, and so in the end because of this difficulty he allowed the name to remain.

The dog had not cost him much to buy, but as it was only a few weeks old he had to keep it at his own cost for fully six months before beginning the business of breaking it, which would take from three to six months longer. A dog cannot be put to work before he is quite half a year old unless he is exceptionally vigorous. Sheep are timid creatures, but not unintelligent, and they can distinguish between the seasoned old sheep-dog, whose furious onset and bite they fear, and the raw young recruit as easily as the rook can distinguish between the man with a gun and the man of straw with a broomstick under his arm. They will turn upon and attack the young dog, and chase him away with his tail between his legs. He will also work too furiously

167

for his strength and then collapse, with the result that he will make a
cowardly sheep-dog, or, as the shepherds say, 'broken-hearted.'

Another thing. He must be made to work at first with an old sheep-
dog, for though he has the impulse to fly about and do something, he
does not know what to do and does not understand his master's gestures and commands. He must have an object-lesson, he must see the
motion and hear the word and mark how the old dog flies to this or
that point and what he does. The word of command or the gesture
thus becomes associated in his mind with a particular action on his
part. But he must not be given too many object-lessons or he will lose
more than he will gain—a something which might almost be described
as a sense of individual responsibility. That is to say, responsibility to
the human master who delegates his power to him. Instead of taking
his power directly from the man he takes it from the dog, and this
becomes a fixed habit so quickly that many shepherds say that if you
give more than from three to six lessons of this kind to a young dog
you will spoil him. He will need the mastership of the other dog, and
will thereafter always be at a loss and work in an uncertain way.

A timid or unwilling young dog is often coupled with the old dog
two or three times, but this method has its dangers too, as it may be
too much for the young dog's strength, and give him that 'broken-
heart' from which he will never recover; he will never be a good
sheep-dog.

To return to Tory. In due time he was trained and proved quick to
learn and willing to work, so that before long he began to be useful
and was much wanted with the sheep, as the old dog was rapidly
growing stiffer on his legs and harder of hearing.

One day the lambs were put into a field which was half clover and
half rape, and it was necessary to keep them on the clover. This the
young dog could not or would not understand; again and again he
allowed the lambs to go to the rape, which so angered Caleb that he
threw his crook at him. Tory turned and gave him a look, then came
very quietly and placed himself behind his master. From that moment
he refused to obey, and Bawcombe, after exhausting all his arts of persuasion, gave it up and did as well as he could without his assistance.

That evening after folding-time he by chance met a shepherd he was
well acquainted with and told him of the trouble he was in over Tory.

'You tie him up for a week,' said the shepherd, 'and treat him well
till he forgets all about it, and he'll be the same as he was before you

offended him. He's just like old Tom—he's got his father's temper.'

'What's that you say?' exclaimed Bawcombe. 'Be you saying that Tory's old Tom's son? I'd never have taken him if I'd known that. Tom's not pure-bred—he's got retriever's blood.'

'Well, 'tis known, and I could have told 'ee, if thee'd asked me,' said the shepherd. 'But you do just as I tell 'ee, and it'll be all right with the dog.'

Tory was accordingly tied up at home and treated well and spoken kindly to and patted on the head, so that there would be no unpleasantness between master and servant, and if he was an intelligent animal he would know that the crook had been thrown not to hurt but merely to express disapproval of his naughtiness.

Then came a busy day for the shepherd, when the lambs were trimmed before being taken to the Wilton sheep-fair. There was Bawcombe, his boy, the decrepit old dog, and Tory to do the work, but when the time came to start Tory refused to do anything.

When sent to turn the lambs he walked off to a distance of about twenty yards, sat down and looked at his master. Caleb hoped he would come round presently when he saw them all at work, and so they did the best they could without him for a time; but the old dog was stiffer and harder of hearing than ever, and as they could not get on properly Caleb went at intervals to Tory and tried to coax him to give them his help; and every time he was spoken to he would get up and come to his master, then when ordered to do something he would walk off to the spot where he had chosen to be and calmly sit down once more and look at them. Caleb was becoming more and more incensed, but he would not show it to the dog; he still hoped against hope; and then a curious thing happened. A swallow came skimming along close to the earth and passed within a yard of Tory, when up jumped the dog and gave chase, darting across the field with such speed that he kept very near the bird until it rose and passed over the hedge at the further side. The joyous chase over Tory came back to his old place, and sitting on his haunches began watching them again struggling with the lambs. It was more than the shepherd could stand; he went deliberately up to the dog, and taking him by the straw collar still on his neck drew him quietly away to the hedge-side and bound him to a bush, then getting a stout stick he came back and gave him one blow on the head. So great was the blow that the dog made not the slightest sound: he fell; his body quivered a moment and his legs stretched out—he was quite dead. Bawcombe then plucked an armful of bracken and threw it over his body to cover it, and going back to the hurdles sent the boy home, then spreading his cloak at the hedge-side, laid himself down on it and covered his head.

An hour later the farmer appeared on the scene. 'What are you doing here, shepherd?' he demanded in surprise. 'Not trimming the lambs!'

Bawcombe, raising himself on his elbow, replied that he was not trimming the lambs—that he would trim no lambs that day.

'Oh, but we must get on with the trimming!' cried the farmer.

Bawcombe returned that the the dog had put him out, and now the dog was dead—he had killed him in his anger, and he would trim no more lambs that day. He had said it and would keep to what he had said.

Then the farmer got angry and said that the dog had a very good nose and would have been useful to him to take rabbits.

'Master,' said the other, 'I got he when he were a pup and broke 'n to help me with the sheep and not to catch rabbits; and now I've killed 'n and he'll catch no rabbits.'

The farmer knew his man, and swallowing his anger walked off without another word.

Later on in the day he was severely blamed by a shepherd friend who said that he could easily have sold the dog to one of the drovers, who were always anxious to pick up a dog in their village, and he would have had the money to repay him for his trouble; to which Bawcombe returned, 'If he wouldn't work for I that broke 'n he wouldn't work for another. But I'll never again break a dog that isn't pure-bred.'

But though he justified himself he had suffered remorse for what he had done; not only at the time, when he covered the dead dog up with bracken and refused to work any more that day, but the feeling had persisted all his life, and he could not relate the incident without showing it very plainly. He bitterly blamed himself for having taken the pup and for spending long months in training him without having first taken pains to inform himself that there was no bad blood in him. And although the dog was perhaps unfit to live he had finally killed him in anger. If it had not been for that sudden impetuous chase after a swallow he would have borne with him and considered afterwards what was to be done; but that dash after the bird was more than he could stand; for it looked as if Tory had done it purposely, in something of a mocking spirit, to exhibit his wonderful activity and speed to his master, sweating there at his task, and make him see what he had lost in offending him.

The shepherd gave another instance of a mistake he once made which caused him a good deal of pain. It was the case of a dog named Bob which he owned when a young man. He was an exceptionally small dog, but his quick intelligence made up for lack of strength, and he was of very lively disposition, so that he was a good companion to a shepherd as well as a good servant.

One summer day at noon Caleb was going to his flock in the fields, walking by a hedge, when he noticed Bob sniffing suspiciously at the roots of an old holly-tree growing on the bank. It was a low but very old tree with a thick trunk, rotten and hollow inside, the cavity being hidden with the brushwood growing up from the roots. As he came abreast of the tree, Bob looked up and emitted a low whine, that sound which says so much when used by a dog to his master and which his master does not always rightly understand. At all events he did not do so in this case. It was August and the shooting had begun, and Caleb jumped to the conclusion that a wounded bird had crept into the hollow tree to hide, and so to Bob's whine, which expressed fear and asked what he was to do, the shepherd answered, 'Get him.' Bob dashed in, but quickly recoiled, whining in a piteous way, and began rubbing his face on his legs. Bawcombe in alarm jumped down and peered into the hollow trunk and heard a slight rustling of dead leaves, but saw nothing. His dog had been bitten by an adder, and he at once returned to the village, bitterly blaming himself for the mistake he had made and greatly fearing that he would lose his dog. Arrived at the village his mother at once went off to the down to inform Isaac of the trouble and ask him what they were to do. Caleb had to wait some time, as none of the villagers who gathered round could suggest a remedy, and in the meantime Bob continued rubbing his cheek against his foreleg, twitching and whining with pain; and before long the face and head began to swell on one side, the swelling extending to the nape and downwards to the throat. Presently Isaac himself, full of concern, arrived on the scene, having left his wife in charge of the flock, and at the same time a man from a neighbouring village came riding by and joined the group. The horseman got off and assisted Caleb in holding the dog while Isaac made a number of incisions with his knife in the swollen place and let out some blood, after which they rubbed the wounds and all the swollen part with an oil used for the purpose. The composition of this oil was a secret: it was made by a man in one of the downland villages and sold at eighteen-pence a small bottle; Isaac was a believer in its efficacy, and always kept a bottle hidden away somewhere in his cottage.

Bob recovered in a few days, but the hair fell out from all the part which had been swollen, and he was a curious-looking dog with half his face and head naked until he got his fresh coat, when it grew again. He was as good and active a dog as ever, and lived to a good old age, but one result of the poison he never got over: his bark had changed from a sharp ringing sound to a low and hoarse one. 'He always barked,' said the shepherd, 'like a dog with a sore throat.'

To go back to the subject of training a dog. Once you make a beginning it must be carried through to a finish. You take him at the age of 6 months and the education must be fairly complete when he is a year old. He is then lively, impressionable, exceedingly adaptive; his intelligence at that period is most like man's; but it would be a mistake to think that it will continue so—that to what he learns now in this wonderful half-year, other things may be added by and by as opportunity arises. At a year he has practically got to the end of his capacity to learn. He has lost his human-like receptivity, but what he has been taught will remain with him for the rest of his life. We can hardly say that he remembers it; it is more like what is called 'inherited memory' or 'lapsed intelligence.'

All this is very important to a shepherd, and explains the reason an old head-shepherd had for saying to me that he had never had, and never would have, a dog he had not trained himself. No two men follow precisely the same method in training, and a dog transferred from his trainer to another man is always a little at a loss; method, voice, gestures, personality, are all different; his new master must study him and in a way adapt himself to the dog. The dog is still more at a loss when transferred from one kind of country to another where the sheep are worked in a different manner, and one instance Caleb gave me of this is worth relating. It was, I thought, one of his best dog stories.

His dogs as a rule were bought as pups; occasionally he had had to get a dog already trained, a painful necessity to a shepherd, seeing that the pound or two it costs—the price of an ordinary animal—is a big sum of money to him. And once in his life he got an old trained sheep-dog for nothing. He was young then, and acting as under-shepherd in his native village, when the report came one day that a great circus and menagerie which had been exhibiting in the west was on its way to Salisbury, and would be coming past the village about six o'clock on the following morning. The turnpike was a little over a mile away, and thither Caleb went with half a dozen other young men of the village at about five o'clock to see the show pass, and sat on a gate beside a wood to wait its coming. In due time the long procession of horses and mounted men and women, and gorgeous vans containing lions and tigers and other strange beasts, came by, affording them great admiration and delight. When it had gone on and the last van had disappeared at the turning of the road, they got down from the gate and were about to set out on their way back when a big, shaggy sheep-dog came out of the wood and running to the road began looking up and down in a bewildered way. They had no doubt that he belonged to the circus and had turned aside to hunt a rabbit in the wood; then, thinking the animal would understand them, they shouted to it and waved their arms in the direction the procession had gone. But the dog became frightened, and turning fled back into cover, and they saw no more of it.

Two or three days later it was rumoured that a strange dog had been seen in the neighbourhood of Winterbourne Bishop, in the fields; and women and children going to or coming from outlying cottages and farms had encountered it, sometimes appearing suddenly out of the furze-bushes and staring wildly at them; or they would meet him in some deep lane between hedges, and after standing still a moment eyeing them he would turn and fly in terror from their strange faces. Shepherds began to be alarmed for the safety of their sheep, and there was a good deal of excitement and talk about the strange dog. Two or three days later Caleb encountered it. He was returning from his flock at the side of a large grass field where four or five women were occupied cutting the thistles, and the dog, which he immediately recognized as the one he had seen at the turnpike, was following one of the women about. She was greatly alarmed, and called to him, 'Come here, Caleb, for goodness' sake, and drive this big dog away! He do look so desprit, I'm afeared of he.'

'Don't you be feared,' he shouted back. 'He won't hurt 'ee; he's starving—don't you see his bones sticking out? He's asking to be fed.'

Then going a little nearer he called to her to take hold of the dog by the neck and keep him while he approached. He feared that the dog on seeing him coming would rush away. After a little while she called the dog, but when he went to her she shrank away from him and called out, 'No, I daren't touch he—he'll tear my hand off. I never see'd such a desprit-looking beast!'

"Tis hunger,' repeated Caleb, and then very slowly and cautiously he approached, the dog all the time eyeing him suspiciously, ready to rush away on the slightest alarm. And while approaching him he began to speak gently to him, then coming to a stand stooped and patting his legs called the dog to him. Presently he came, sinking his body lower as he advanced and at last crawling, and when he arrived at the shepherd's feet he turned himself over on his back—that eloquent action which a dog uses when humbling himself before and imploring mercy from one mightier than himself, man or dog.

Caleb stooped, and after patting the dog gripped him firmly by the neck and pulled him up, while with his free hand he undid his leather belt to turn it into a dog's collar and leash; then, the end of the strap in his hand, he said 'Come,' and started home with the dog at his side. Arrived at the cottage he got a bucket and mixed as much meal as would make two good feeds, the dog all the time watching him with his muscles twitching and the water running from his mouth. The meal well mixed he emptied it out on the turf, and what followed, he said, was an amazing thing to see: the dog hurled himself down on the food and started devouring it as if the mass of meal had been some living savage creature he had captured and was frenziedly tearing to pieces. He turned round and round, floundering on the earth, uttering strange noises like half-choking growls and screams while gobbling down the meal; then when he had devoured it all he began tearing up and swallowing the turf for the sake of the little wet meal still adhering to it.

Such rage of hunger Caleb had never seen, and it was painful to him to think of what the dog had endured during those days when it had been roaming foodless about the neighbourhood. Yet it was among sheep all the time—scores of flocks left folded by night at a distance from the village; one would have imagined that the old wolf and wild-dog instinct would have come to life in such circumstances, but the instinct was to all appearance dead.

My belief is that the pure-bred sheep-dog is indeed the last dog to re-

174

vert to a state of nature; and that when sheep-killing by night is traced to a sheep-dog, the animal has a bad strain in him, of retriever, or cur, or 'rabbit-dog,' as the shepherds call all terriers. When I was a boy on the pampas sheep-killing dogs were common enough, and they were always curs, or the common dog of the country, a smooth-haired animal about the size of a coach-dog, red, or black, or white. I recall one instance of sheep-killing being traced to our own dogs—we had about six or eight just then. A native neighbour, a few miles away, caught them at it one morning; they escaped him in spite of his good horse, with lasso and bolas also, but his sharp eyes saw them pretty well in the dim light, and by and by he identified them, and my father had to pay him for about thirty slain and badly injured sheep; after which a gallows was erected and our guardians ignominiously hanged. Here we shoot dogs; in some countries the old custom of hanging them, which is perhaps less painful, is still followed.

It was common, too, in those days on the pampas, especially in the outlying districts, for dogs to take to a wholly wild life. I remember once, when staying with a native friend among the Sierras, near Cape Corrientos, that he owned a fine handsome dog, so good-tempered and intelligent that I was very much attached to him. He was, my friend said, a wild dog; he had found a bitch with a litter of pups in a huge burrow she had made for herself in the ground, and he had killed them all except this one, which he took home and reared as an experiment.

In England it is perhaps now impossible for a dog to run wholly wild, or to exist in that state for any length of time. I find one case reported in the 'Salisbury Journal' of 31 May, 1779. It interested me very much because I had long been familiar with the place in which the escaped dog was found. This was Pamba Wood, near Silchester, and is sometimes called Silchester Forest – the 'Proud Pamba' of Drayton's 'Polyolbion'.

A poor woman while gathering sticks in the wood came upon the remains of a dead man—the skull and a number of bones. She gave notice of it, and a crowd of villagers went to the spot, and found there a foxhound bitch which had been missing from the kennels for about two months. She had a litter of eight pups about two months old in a pit about six feet deep which she had made herself, and it was plain that the dogs had devoured the flesh of the man after he had met his death close to the pit. Nothing except the flesh on his feet and ankles re-

175

mained uneaten: they were cased in thick high boots, so hard that the dogs had been unable to tear them to pieces. The dead man was identified or taken to be a thrasher from the neighbouring village of Aldermaston who had been missing about a fortnight, and it was supposed that he had gone into the wood to cut a flail and had been seized with sudden illness and died at that spot.

The pups were very wild and savage, and I should rather think that the man had found the bitch with her young and had attempted to take them singlehanded for the sake of the reward, and had been attacked, pulled down, and killed, after which they began to feed on the body. One wonders how this dog had managed to support herself and her eight pups in the forest during the six weeks before the poor thrasher came in their way to provide them with food.

In this same journal I find a case of a dog devouring its own master. The man was a rat-catcher living in a cottage at Fovant, a small village in the valley of the Nadder. Going home drunk one night from the public-house he fell in the road, and remained lying there in a drunken sleep, and towards morning a wagon passed over him, the wheel crushing his head. The wagoner reported the case, and the constable with men to help him went and removed the body to the man's cottage, and after depositing it on the floor of the kitchen or living room went their way, closing the door after them. Later in the day they returned, and going in found the dog devouring the man's flesh.

We experience an intense disgust at a case like this; but we have the same feeling when we hear of man eating dog; in both cases, owing to the long and intimate association between man and dog, we are affected as by a kind of cannibalism.

To go back to our story. From that time the stray dog was Caleb's obedient and affectionate slave, always watching his face and every gesture, and starting up at his slightest word in readiness to do his bidding. When put with the flock he turned out to be a useful sheep-dog, but unfortunately he had not been trained on the Wiltshire Downs. It was plain to see that the work was strange to him, that he had been taught in a different school, and could never forget the old and acquire a new method. But as to what conditions he had been reared in or in what district or country no one could guess. Every one said that he was a sheep-dog, but unlike any sheep-dog they had ever seen; he was not Wiltshire, nor Welsh, nor Sussex, nor Scotch, and they could say no more. Whenever a shepherd saw him for the first time his attention

176

was immediately attracted, and he would stop to speak with Caleb. 'What sort of a dog do you call that?' he would say. 'I never see'd one just like 'n before.'

At length one day when passing by a new building which some workmen had been brought from a distance to erect in the village, one of the men hailed Caleb and said, 'Where did you get that dog, mate?'

'Why do you ask me that?' said the shepherd.

'Because I know where he come from: he's a Rooshian, that's what he is. I've see'd many just like him in the Crimea when I was there. But I never see'd one before in England.'

Caleb was quite ready to believe it, and was a little proud at having a sheep-dog from that distant country. He said that it also put something new into his mind. He didn't know nothing about Russia before that, though he had been hearing so much of our great war there and of all the people that had been killed. Now he realized that Russia was a great country, a land where there were hills and valleys and villages, where there were flocks and herds, and shepherds and sheep-dogs just as in the Wiltshire Downs. He only wished that Tramp—that was the name he had given his dog—could have told him his history.

Tramp, in spite of being strange to the downs and the downland sheep-dog's work, would probably have been kept by Caleb to the end but for his ineradicable passion for hunting rabbits. He did not neglect his duty, but he would slip away too often, and eventually when a man who wanted a good dog for rabbits one day offered Caleb fifteen shillings for Tramp, he sold him, and as he was taken away to a distance by his new master, he never saw him again.

XXII

THE SHEPHERD AS NATURALIST

General remarks—Great Ridge Wood—Encounter with a roe-deer—A hare on a stump—A gamekeeper's memory—Talk with a gipsy—A strange story of a hedgehog—A gipsy on memory—The shepherd's feeling for animals—Anecdote of a shrew—Anecdote of an owl—Reflex effect of the gamekeeper's calling—We remember best what we see emotionally

IT WILL appear to some of my readers that the interesting facts about wild life, or rather about animal life, wild and domestic, gathered in my talks with the old shepherd, do not amount to much. If this is all there is to show after a long life spent out of doors, or all that is best worth preserving, it is a somewhat scanty harvest, they will say. To me it appears a somewhat abundant one. We field naturalists, who set down what we see and hear in a notebook lest we forget it, do not always bear in mind that it is exceedingly rare for those who are not naturalists, whose senses and minds are occupied with other things, to come upon a new and interesting fact in animal life, or that these chance observations are quickly forgotten. This was strongly borne in upon me lately while staying in the village of Hindon in the neighbourhood of the Great Ridge Wood, which clothes the summit of the long high down overlooking the vale of the Wylye. It is an immense wood, mostly of scrub or dwarf oak, very dense in some parts, in others thin, with open, barren patches, and like a wild forest, covering altogether twelve or fourteen square miles—perhaps more. There are no houses near, and no people in it except a few gamekeepers: I spent long days in it without meeting a human being. It was a joy to me to find such a spot in England, so wild and solitary, and I was filled with pleasing anticipation of all the wild life I should see in such a place, especially after an experience I had on my second day in it. I was standing in an open glade when a cock-pheasant uttered a cry of alarm, and immediately afterwards, startled by the cry perhaps, a roe-deer rushed out of the close thicket of oak and holly in which it had been hiding, and ran past me at a very short distance, giving me a good sight of this shyest of the large wild animals still left to us. He looked very beautiful to me, in that mouse-coloured coat which makes him invisible in the deep shade in which he is accustomed to pass the daylight hours in hiding, as he fled across the green open space in the brilliant May sunshine. But he was only one, a chance visitor, a wanderer from wood to wood about the land; and he had been seen once, a month before my encounter with him, and ever since then the keepers

had been watching and waiting for him, gun in hand, to send a charge of shot into his side.

That was the best and the only great thing I saw in the Great Ridge Wood, for the curse of the pheasant is on it as on all woods and forests in Wiltshire, and all wild life considered injurious to the semi-domestic bird, from the sparrowhawk to the harrier and buzzard and goshawk, and from the little mousing weasel to the badger; and all the wild life that is only beautiful, or which delights us because of its wildness, from the squirrel to the roe-deer, must be included in the slaughter.

One very long summer day spent in roaming about in this endless wood, always on the watch, had for sole result, so far as anything out of the common goes, the spectacle of a hare sitting on a stump. The hare started up at a distance of over a hundred yards before me and rushed straight away at first, then turned and ran on my left so as to get round to the side from which I had come. I stood still and watched him as he moved swiftly over the ground, seeing him not as a hare but as a dim brown object successively appearing, vanishing, and reappearing, behind and between the brown tree-trunks, until he had traced half a circle and was then suddenly lost to sight. Thinking that he had come to a stand I put my binocular on the spot where he had vanished, and saw him sitting on an old oak stump about thirty inches high.

179

It was a round mossy stump about eighteen inches in diameter, standing in a bed of brown dead leaves, with the rough brown trunks of other dwarf oak-trees on either side of it. The animal was sitting motionless, in profile, its ears erect, seeing me with one eye, and was like a carved figure of a hare set on a pedestal, and had a very striking appearance.

As I had never seen such a thing before I thought it was worth mentioning to a keeper I called to see at his lodge on my way back in the evening. It had been a blank day, I told him—a hare sitting on a stump being the only thing I could remember to tell him. 'Well,' he said, 'you've seen something I've never seen in all the years I've been in these woods. And yet, when you come to think of it, it's just what one might expect a hare would do. The wood is full of old stumps, and it seems only natural a hare should jump on to one to get a better view of a man or animal at a distance among the trees. But I never saw it.'

What, then, had he seen worth remembering during his long hours in the wood on that day, or the day before, or on any day during the last thirty years since he had been policing that wood, I asked him. He answered that he had seen many strange things, but he was not now able to remember one to tell me! He said, further, that the only things he remembered were those that related to his business of guarding and rearing the birds; all other things he observed in animals, however remarkable they might seem to him at the moment, were things that didn't matter and were quickly forgotten.

On the very next day I was out on the down with a gipsy, and we got talking about wild animals. He was a middle-aged man and a very perfect specimen of his race—not one of the blue-eyed and red or light-haired bastard gipsies, but dark as a Red Indian, with eyes like a hawk, and altogether a hawk-like being, lean, wiry, alert, a perfectly wild man in a tame, civilized land. The lean, mouse-coloured lurcher that followed at his heels was perfect too, in his way—man and dog appeared made for one another. When this man spoke of his life, spent in roaming about the country, of his very perfect health, and of his hatred of houses, the very atmosphere of any indoor place producing a suffocating and sickening effect on him, I envied him as I envy birds their wings and as I can never envy men who live in mansions. His was the wild, the real life, and it seemed to me that there was no other worth living.

'You know,' said he, in the course of our talk about wild animals, 'we are very fond of hedgehogs—we like them better than rabbits.'

'Well, so do I,' was my remark. I am not quite sure that I do, but that is what I told him. 'But now you talk of hedgehogs,' I said, 'it's funny to think that, common as the animal is, it has some queer habits I can't find anything about from gamekeepers and others I've talked to on the subject, or from my own observation. Yet one would imagine that we know all there is to be known about the little beast; you'll find his history in a hundred books—perhaps in five hundred. There's one book about our British animals so big you'd hardly be able to lift its three volumes from the ground with all your strength, in which its author has raked together everything known about the hedgehog, but he doesn't give me the information I want—just what I went to the book to find. Now here's what a friend of mine once saw. He's not a naturalist, nor a sportsman, nor a gamekeeper, and not a gipsy; he doesn't observe animals or want to find out their ways; he is a writer, occupied day and night with his writing, sitting among books, yet he saw something which the naturalists and gamekeepers haven't seen, so far as I know. He was going home one moonlight night by a footpath through the woods when he heard a very strange noise a little distance ahead, a low whistling sound, very sharp, like the continuous twittering of a little bird with a voice like a bat, or a shrew, only softer, more musical. He went on very cautiously, until he spied two hedgehogs standing on the path facing each other, with their noses almost or quite touching. He remained watching and listening to them for some moments, then tried to go a little nearer and they ran away.

'Now I've asked about a dozen gamekeepers if they ever saw such a thing, and all said they hadn't; they never heard hedgehogs make that twittering sound, like a bird or a singing mouse; they had only heard them scream like a rabbit when in a trap. Now what do you say about it?'

'I've never seen anything like that,' said the gipsy. 'I only know the hedgehog makes a little whistling sound when he first comes out at night; I believe it is a sort of call they have.'

'But no doubt,' I said, 'you've seen other queer things in hedgehogs and in other little animals which I should like to hear.'

Yes, he had, first and last, seen a good many queer things both by day and night, in woods and other places, he replied, and then continued; 'But you see it's like this. We see something and say, "Now that's a very curious thing!" and then we forget all about it. You see, we don't lay no store by such things; we ain't scholards and don't know nothing about what's said in books. We see something and say *That's* something we never saw before and never heard tell of, but maybe others have seen it and you can find it in the books. So that's how 'tis, but if I hadn't forgotten them I could have told you a lot of queer things.'

That was all he could say, and few can say more. Caleb was one of the few who could, and one wonders why it was so, seeing that he was occupied with his own tasks in the fields and on the down where wild life is least abundant and varied, and that his opportunities were so few compared with those of the game-keeper. It was, I take it, because he had sympathy for the creatures he observed, that their actions had stamped themselves on his memory, because he had seen them emotionally. We have seen how well he remembered the many sheep-dogs he had owned, how vividly their various characters are portrayed in his account of them. I have met with shepherds who had little to tell about the dogs they had possessed; they had regarded their dogs as useful servants and nothing more as long as they lived, and when dead they were forgotten. But Caleb had a feeling for his dogs which made it impossible for him to forget them or to recall them without that tenderness which accompanies the thought of vanished human friends. In a lesser degree he had something of this feeling for all animals, down even to the most minute and unconsidered. I recall here one of his anecdotes of a very small creature—a shrew, or over-runner, as he called it.

One day when out with his flock a sudden storm of rain caused him to seek for shelter in an old untrimmed hedge close by. He crept into the ditch, full of old dead leaves beneath the tangle of thorns and brambles, and setting his back against the bank he thrust his legs out, and as he did so was startled by an outburst of shrill little screams at his feet. Looking down he spied a shrew standing on the dead leaves close to his boot, screaming with all its might, its long thin snout pointed upwards and its mouth wide open; and just above it, two or three inches perhaps, hovered a small brown butterfly. There for a few moments it continued hovering while the shrew continued screaming; then the butterfly flitted away and the shrew disappeared among the dead leaves.

Caleb laughed (a rare thing with him) when he narrated this little incident, then remarked: 'The over-runner was a-crying 'cause he couldn't catch that leetel butterfly.'

The shepherd's inference was wrong; he did not know—few do—that the shrew has the singular habit, when surprised on the surface and in danger, of remaining motionless and uttering shrill cries. His foot, set down close to it, had set it screaming; the small butterfly, no doubt disturbed at the same moment, was there by chance. I recall here another little story he related of a bird—a long-eared owl.

One summer there was a great drought, and the rooks, unable to get their usual food from the hard, sun-baked pasture-lands, attacked the roots and would have pretty well destroyed them if the farmer had not protected his swedes by driving in stakes and running lines of cotton-thread and twine from stake to stake all over the field. This kept them off, just as thread keeps the chaffinches from the seed-beds in small gardens, and as it keeps the sparrows from the crocuses on lawns and ornamental grounds. One day Caleb caught sight of an odd-looking, brownish-grey object out in the middle of the turnip-field, and as he looked it rose up two or three feet into the air, then dropped back again, and this curious movement was repeated at intervals of two or three minutes until he went to see what the thing was. It turned out to be a long-eared owl, with its foot accidentally caught by a slack thread, which allowed the bird to rise a couple of feet into the air; but every such attempt to escape ended in its being pulled back to the ground again. It was so excessively lean, so weightless in his hand, when he took it up after disengaging its foot, that he thought it must have been captive for the space of two or three days. The wonder was that it had kept alive during those long midsummer days of intolerable heat out there in the middle of the burning field. Yet it was in very fine feather and beautiful to look at with its long, black ear-tufts and round, orange-yellow eyes, which would never lose their fiery lustre until glazed in death. Caleb's first thought on seeing it closely was that it would have been a prize to anyone who liked to have a handsome bird stuffed in a glass case. Then raising it over his head he allowed it to fly, whereupon it flew off a distance of a dozen or fifteen yards and pitched among the turnips, after which it ran a little space and rose again with labour, but soon recovering strength it flew away over the field and finally disappeared in the deep shade of the copse beyond.

In relating these things the voice, the manner, the expression in his eyes, were more than the mere words, and displayed the feeling which had caused these little incidents to endure so long in his memory.

The gamekeeper cannot have this feeling: he may come to his task

with the liveliest interest in, even with sympathy for, the wild creatures amidst which he will spend his life, but it is all soon lost. His
business in the woods is to kill, and the reflex effect is to extinguish all
interest in the living animal—in its life and mind. It would, indeed, be
a wonderful thing if he could remember any singular action or appearance of an animal which he had witnessed before bringing his gun
automatically to his shoulder.

XXIII

THE MASTER OF THE VILLAGE

Moral effect of the great man—An orphaned village—The masters of the village—Elijah Raven—Strange appearance and character—Elijah's house —The owls—Two rooms in the house—Elijah hardens with time—The village club and its arbitrary secretary—Caleb dips the lambs and falls ill —His claim on the club rejected—Elijah in court

IN MY roamings about the downs it is always a relief—a positive pleasure in fact—to find myself in a village which has no squire or other magnificent and munificent person who dominates everybody and everything, and, if he chooses to do so, plays providence in the community. I may have no personal objection to him—he is sometimes almost if not quite human; what I heartily dislike is the effect of his position (that of a giant among pigmies) on the lowly minds about him, and the servility, hypocrisy, and parasitism which spring up and flourish in his wide shadow whether he likes these moral weeds or not. As a rule he likes them, since the poor devil has this in common with the rest of us, that he likes to stand high in the general regard. But how is he to know it unless he witnesses its outward beautiful signs every day and every hour on every countenance he looks upon? Better, to my mind, the severer conditions, the poverty and unmerited sufferings which cannot be relieved, with the greater manliness and self-dependence when the people are left to work out their own destiny. On this account I was pleased to make the discovery on my first visit to Caleb's native village that there was no magnate, or other big man, and no gentleman except the parson, who was not a rich man. It was, so to speak, one of the orphaned villages left to fend for itself and fight its own way in a hard world, and had nobody even to give the customary blankets and sack of coals to its old women. Nor was there any very big farmer in the place, certainly no gentleman farmer; they were mostly small men, some of them hardly to be distinguished in speech and appearance from their hired labourers.

In these small isolated communities it is common to find men who have succeeded in rising above the others and in establishing a sort of mastery over them. They are not as a rule much more intelligent than

the others who are never able to better themselves; the main difference is that they are harder and more grasping and have more self-
control. These qualities tell eventually, and set a man a little apart, a
little higher than the others, and he gets the taste of power, which re-
acts on him like the first taste of blood on the big cat. Henceforward
he has his ideal, his definite goal, which is to get the upper hand—to
be on top. He may be, and generally is, an exceedingly unpleasant fellow to have for a neighbour—mean, sordid, greedy, tyrannous, even
cruel, and he may be generally hated and despised as well, but along
with these feelings there will be a kind of shamefaced respect and
admiration for his courage in following his own line in defiance of
what others think and feel. It is after all with man as with the social
animals: he must have a master—not a policeman, or magistrate, or a
vague, far-away, impersonal something called the authorities or the
government; but a head of the pack or herd, a being like himself
whom he knows and sees and hears and feels every day. A real man,
dressed in old familiar clothes, a fellow-villager, who, wolf or dog-like,
has fought his way to the mastership.

There was a person of this kind at Winterbourne Bishop who was
often mentioned in Caleb's reminiscences, for he had left a very
strong impression on the shepherd's mind—as strong, perhaps, though
in a disagreeable way, as that of Isaac his father, and of Mr. Ellerby of
Doveton. For not only was he a man of great force of character, but
he was of eccentric habits and of a somewhat grotesque appearance.

The curious name of this person was Elijah Raven. He was a
native of the village and lived till extreme old age in it, the last of his
family, in a small house inherited from his father, situated about the
centre of the village street. It was a quaint, old, timbered house, little
bigger than a cottage, with a thatched roof, and behind it some out-
buildings, a small orchard, and field of a dozen or fifteen acres. Here
he lived with one other person, an old man who did the cooking and
housework, but after this man died he lived alone. Not only was he a
bachelor, but he would never allow any woman to come inside his
house. Elijah's one idea was to get the advantage of others—to make
himself master in the village. Beginning poor, he worked in a small,
cautious, peddling way at farming, taking a field or meadow or strip
of down here and there in the neighbourhood, keeping a few sheep, a
few cows, buying and selling and breeding horses. The men he
employed were those he could get at low wages—poor labourers who

were without a place and wanted to fill up a vacant time, or men like the Targetts described in a former chapter who could be imposed upon; also gipsies who flitted about the country, working in a spasmodic way when in the mood for the farmers who could tolerate them, and who were paid about half the wages of an ordinary labourer. If a poor man had to find money quickly, on account of illness or some other cause, he could get it from Elijah at once—not borrowed, since Elijah neither lent nor gave—but he could sell him anything he possessed—a horse or cow, or sheep-dog, or a piece of furniture; and if he had nothing to sell, Elijah would give him something to do and pay him something for it. The great thing was that Elijah had money which he was always willing to circulate. At his unlamented death he left several thousands of pounds, which went to a distant relation, and a name which does not smell sweet, but is still remembered not only at Winterbourne Bishop but at many other villages on Salisbury Plain.

Elijah was short of stature, broad shouldered, with an abnormally big head and large dark eyes. They say that he never cut his hair in his life. It was abundant and curly, and grew to his shoulders, and when he was old and his great mass of hair and beard became white it was said that he resembled a gigantic white owl. Mothers frightened their children into quiet by saying 'Elijah will get you if you don't behave yourself.' He knew and resented this, and though he never noticed a child he hated to have the little ones staring in a half-terrified way at him. To seclude himself more from the villagers he planted holly and yew bushes before his house, and eventually the entire building was hidden from sight by the dense evergreen thicket. The trees were cut down after his death: they were gone when I first visited the village and by chance found a lodging in the house, and congratulated myself that I had got the quaintest, old rambling rooms I had ever inhabited. I did not know that I was in Elijah Raven's house, although his name had long been familiar to me: it only came out one day when I asked my landlady, who was a native, to tell me the history of the place. She remembered how as a little girl, full of mischief and greatly daring, she had sometimes climbed over the low front wall to hide under the thick yew bushes and watch to catch a sight of the owlish old man at his door or window.

For many years Elijah had two feathered tenants, a pair of white owls —the birds he so much resembled. They occupied a small garret at the end of his bedroom, having access to it through a hole under the thatch.

They bred there in peace and on summer evenings one of the common sights of the village was Elijah's owls flying from the house behind the evergreens and returning to it with mice in their talons. At such seasons the threat to the unruly children would be varied to 'Old Elijah's owls will get you.' Naturally, the children grew up with the idea of the birds and the owlish old man associated in their minds.

It was odd that the two very rooms which Elijah had occupied during all those solitary years, the others being given over to spiders and dust, should have been assigned to me when I came to lodge in the house. The first, my sitting-room, was so low that my hair touched the ceiling when I stood up my full height; it had a brick floor and a wide old fireplace on one side. Though so low-ceilinged it was very large and good to be in when I returned from a long ramble on the downs, sometimes wet and cold, to sit by a wood fire and warm myself. At night when I climbed to my bedroom by means of the narrow, crooked, worm-eaten staircase, with two difficult and dangerous corners to get round, I would lie awake staring at the small square patch of greyness in the black interior made by the latticed window; and listening to the wind and rain outside, would remember that the sordid, owlish old man had slept there and stared nightly at that same grey patch in the dark for very many years. If, I thought, that something of a man which remains here below to haunt the scene of its past life is more likely to exist and appear to mortal eyes in the case of a person of strong individuality, then there is a chance that I may be visited this night by Elijah Raven his ghost. But his owlish countenance never appeared between me and that patch of pale dim light; nor did I ever feel a breath of cold unearthly air on me.

Elijah did not improve with time; the years that made him long-haired, whiter, and more owl-like also made him more penurious and grasping, and anxious to get the better of every person about him. There was scarcely a poor person in the village—not a field labourer nor shepherd nor farmer's boy, nor any old woman he had employed, who did not consider that they had suffered at his hands. The very poorest could not escape; if he got some one to work for fourpence a day he would find a reason to keep back a portion of the small sum due to him. At the same time he wanted to be well thought of, and at length an opportunity came to him to figure as one who did not live wholly for himself but rather as a person ready to go out of his way to help his neighbours.

There had long existed a small benefit society or club in the village to which most of the farm-hands in the parish belonged, the members numbering about sixty or seventy. Subscriptions were paid quarterly, but the rules were not strict, and any member could take a week or a fortnight longer to pay; when a member fell ill he received half the amount of his wages a week from the funds in hand, and once a year they had a dinner. The secretary was a labourer and in time he grew old and infirm and could not hold a pen in his rheumaticky fingers, and a meeting was held to consider what was to be done in the matter. It was not an easy one to settle. There were few members capable of keeping the books who would undertake the duty, as it was unpaid, and no one among them well known and trusted by all the members. It was then that Elijah Raven came to the rescue. He attended the meeting, which he was allowed to do owing to his being a person of importance—the only one of that description in the village; and getting up on his legs he made the offer to act as secretary himself. This came as a great surprise, and the offer was at once and unanimously accepted, all unpleasant feelings being forgotten, and for the first time in his life Elijah heard himself praised as a disinterested person, one it was good to have in the village.

Things went on very well for a time, and at the yearly dinner of the club, a few months later, Elijah gave an account of his stewardship, showing that the club had a surplus of two hundred pounds. Shortly after this trouble began; Elijah, it was said, was making use of his position as secretary for his own private interests and to pay off old scores against those he disliked. When a man came with his quarterly subscription Elijah would perhaps remember that this person had refused to work for him or that he had some quarrel with him, and if the subscription was overdue he would refuse to take it; he would tell the man that he was no longer a member, and he also refused to give sick pay to any applicant whose last subscription was still due, if he happened to be in Elijah's black book. By and by he came into collision with Caleb, one of the villagers against whom he cherished a special grudge, and this small affair resulted in the dissolution of the club.

At this time Caleb was head-shepherd at Bartle's Cross, a large farm above a mile and a half from the village. One excessively hot day in August he had to dip the lambs; it was very hard work to drive them from the farm over a high down to the stream a mile below the village, where there was a dipping place, and he was tired and hot, and in a sweat when he began the work. With his arms bared to the shoulders he took and plunged his first lamb into the tank. When engaged in dipping, he said, he always kept his mouth closed tightly for fear of getting even a drop of the mixture in it, but on this occasion it unfortunately happened that the man assisting him spoke to him and he was compelled to reply, but had no sooner opened his mouth to speak than the lamb made a violent struggle in his arms and splashed the water over his face and into his mouth. He got rid of it as quickly as he could, but soon began to feel bad, and before the work was over he had to sit down two or three times to rest. However, he struggled on to the finish, then took the flock home and went to his cottage. He could do no more. The farmer came to see what the matter was, and found him in a fever, with face and throat greatly swollen. 'You look bad,' he said; 'you must be off to the doctor.' But it was five miles to the village where the doctor lived, and Bawcombe replied that he couldn't go. 'I'm too bad—I couldn't go, master, if you offered me money for it,' he said.

Then the farmer mounted his horse and went himself, and the doctor came. 'No doubt,' he said, 'you've got some of the poison into your system and took a chill at the same time.' The illness lasted six weeks, and then the shepherd resumed work, although still feeling very shaky. By and by when the opportunity came, he went to claim his sick pay—six shillings a week for the six weeks, his wages being then twelve shillings. Elijah flatly refused to pay him: his subscription, he said, had been due for several weeks and he had consequently forfeited his right to anything. In vain the shepherd explained that he could not pay when lying ill at home with no money in the house and receiving no pay from the farmer. The old man remained obdurate, and with a very heavy heart the shepherd came out and three or four of the villagers waiting in the road outside to hear the result of the application. They, too, were men who had been turned away from the club by the arbitrary secretary. Caleb was telling them about his interview when Elijah came out of the house and leaning over the front gate began to listen. The shepherd then turned towards him and said in a loud voice: 'Mr. Elijah Raven, don't you think this is a tarrible hard case! I've paid my subscription every quarter for thirty years and never had nothing from the fund except two weeks' pay when I were bad some years ago. Now I've been bad six weeks, and my master giv' me nothing for that time, and I've got the doctor to pay and nothing to live on. What am I to do?'

Elijah stared at him in silence for some time, then spoke: 'I told you in there I wouldn't pay you one penny of the money and I'll hold to what I said—in there I said it indoors, and I say again that indoors I'll never pay you—no, not one penny piece. But if I happen some day to meet you out of doors then I'll pay you. Now go.'

And go he did, very meekly, his wrath going down as he trudged home; for after all he would have his money by and by, although the hard old man would punish him for past offences by making him wait for it.

A week or so went by, and then one day while passing through the village he saw Elijah coming towards him, and said to himself, Now I'll be paid! When the two men drew near together he cried out cheerfully, 'Good morning, Mr. Raven.' The other without a word and without a pause passed by on his way, leaving the poor shepherd gazing crestfallen after him.

After all he would not get his money! The question was discussed in the cottages, and by and by one of the villagers who was not so poor as most of them, and went occasionally to Salisbury, said he would ask an attorney's advice about the matter. He would pay for the advice out of his own pocket; he wanted to know if Elijah could lawfully do such things.

To the man's astonishment the attorney said that as the club was not registered and the members had themselves made Elijah their head he could do as he liked—no action would lie against him. But if it was true and it could be proved that he had spoken those words about paying the shepherd his money if he met him out of doors, then he could be made to pay. He also said he would take the case up and bring it into court if a sum of five pounds was guaranteed to cover expenses in case the decision went against them.

Poor Caleb, with twelve shillings a week to pay his debts and live on, could guarantee nothing, but by and by when the lawyer's opinion had been discussed at great length at the inn and in all the cottages in the village, it was found that several of Bawcombe's friends were willing to contribute something towards a guarantee fund, and eventually the sum of five pounds was raised and handed over to the person who had seen the lawyer.

His first step was to send for Bawcombe, who had to get a day off and journey in the carrier's cart one market-day to Salisbury. The result was that action was taken, and in due time the case came on. Elijah Raven was in court with two or three of his friends—small working farmers who had some interested motive in desiring to appear as his supporters. He, too, had engaged a lawyer to conduct his case. The judge, said Bawcombe, who had never seen one before, was a tarrible stern-looking old man in his wig. The plaintiff's lawyer he did open the case and he did talk and talk a lot, but Elijah's counsel he did keep on interrupting him, and they two argued and argued, but the judge he never said no word, only he looked blacker and more tarrible stern. Then when the talk did seem all over, Bawcombe, ignorant of the forms, got up and said, 'I beg your lordship's pardon, but may I speak?' He didn't rightly remember afterwards what he called him, but 'twere your lordship or your worship, he was sure. 'Yes, certainly, you are here to speak,' said the judge, and Bawcombe then gave an account of his interview with Elijah and of the conversation outside the house.

Then up rose Elijah Raven, and in a loud voice exclaimed, 'Lord, Lord, what a sad thing it is to have to sit here and listen to this man's lies!'

'Sit down, sir,' thundered the judge; 'sit down and hold your tongue, or I shall have you removed.'

Then Elijah's lawyer jumped up, and the judge told him he'd better sit down too because he knowed who the liar was in this case. 'A brutal case!' he said, and that was the end, and Bawcombe got his six weeks' sick pay and expenses, and about three pounds besides, being his share of the society's funds which Elijah had been advised to distribute to the members.

And that was the end of the Winterbourne Bishop club, and from that time it has continued without one.

XXIV

ISAAC'S CHILDREN

Isaac Bawcombe's family—The youngest son—Caleb goes to seek David at Wilton sheep-fair—Martha, the eldest daughter—Her beauty—She marries Shepherd Ierat—The name of Ierat—Story of Ellen Ierat—The Ierats go to Somerset—Martha and the lady of the manor—Martha's travels—Her mistress dies—Return to Winterbourne Bishop—Shepherd Ierat's end

CALEB WAS one of five, the middle one, with a brother and sister older and a brother and sister younger than himself—a symmetrical family. I have already written incidentally of the elder brother and the youngest sister, and in this chapter will complete the history of Isaac's children by giving an account of the eldest sister and youngest brother.

The brother was David, the hot-tempered young shepherd who killed his dog Monk, and who afterwards followed his brother to Warminster. In spite of his temper and 'want of sense' Caleb was deeply attached to him, and when as an old man his shepherding days were finished he followed his wife to their new home, he grieved at being so far removed from his favourite brother. For some time he managed to make the journey to visit him once a year. Not to his home near Warminster, but to Wilton, at the time of the great annual sheep-fair held on 12 September. From his cottage he would go by the carrier's cart to the nearest town, and thence by rail with one or two changes by Salisbury and Wilton.

After I became acquainted with Caleb he was ill and not likely to recover, and for over two years could not get about. During all this time he spoke often to me of his brother and wished he could see him. I wondered why he did not write; but he would not, nor would the other. These people of the older generation do not write to each other; years are allowed to pass without tidings, and they wonder and wish and talk of this and that absent member of the family, trusting it is well with them, but to write a letter never enters into their minds.

At last Caleb began to mend and determined to go again to Wilton sheep-fair to look for his beloved brother; to Warminster he could not go; it was too far. September the 12th saw him once more at the old meeting-place, painfully making his slow way to that part of the

193

ground where Shepherd David Bawcombe was accustomed to put his sheep. But he was not there. 'I be here too soon,' said Caleb, and sat himself patiently down to wait, but hours passed and David did not appear, so he got up and made his way about the fair in search of him, but couldn't find 'n. Returning to the old spot he got into conversation with two young shepherds and told them he was waiting for his brother who always put his sheep in that part. 'What be his name?' they asked, and when he gave it they looked at one another and were silent. Then one of them said, 'Be you Shepherd Caleb Bawcombe?' and when he had answered them the other said, 'You'll not see your brother at Wilton today. We've come from Doveton, and knew he. You'll not see your brother no more. He be dead these two years.'

Caleb thanked them for telling him, and got up and went his way very quietly, and got back that night to his cottage. He was very tired, said his wife; he wouldn't eat and he wouldn't talk. Many days passed and he still sat in his corner and brooded, until the wife was angry and said she never knowed a man make so great a trouble over losing a brother. 'Twas not like losing a wife or a son, she said; but he answered not a word, and it was many weeks before that dreadful sadness began to wear off, and he could talk cheerfully once more of his old life in the village.

Of the sister, Martha, there is much more to say; her life was an eventful one as lives go in this quiet downland country, and she was, moreover, distinguished above the others of the family by her beauty and vivacity. I only knew her when her age was over 80, in her native village where her life ended some time ago, but even at that age there was something of her beauty left and a good deal of her charm. She had a good figure still and was of a good height; and had dark, fine eyes, clear, dark, unwrinkled skin, a finely shaped face, and her grey hair, once black, was very abundant. Her manner, too, was very engaging. At the age of 25 she married a shepherd named Thomas Ierat—a surname I had not heard before and which made me wonder where were the Ierats in Wiltshire that in all my rambles among the downland villages I had never come across them, not even in the churchyards. Nobody knew—there ware no Ierats except Martha Ierat, the widow, of Winterbourne Bishop and her son—nobody had ever heard of any other family of the name. I began to doubt that there ever had been such a name until quite recently when, on going over an old downland village church, the rector took me out to show me 'a strange name' on a tablet let into the wall of the building outside. The name was Ierat and the date the seventeenth century. He had never seen the name excepting on that tablet. Who, then, was Martha's husband? It was a queer story which she would never have told me, but I had it from her brother and his wife.

A generation before that of Martha, at a farm in the village of Bower Chalk on the Ebble, there was a girl named Ellen Ierat employed as a dairymaid. She was not a native of the village, and if her parentage and place of birth were ever known they have long passed out of memory. She was a good-looking, nice-tempered girl, and was much liked by her master and mistress, so that after she had been about two years in their service it came as a great shock to find that she was in the family way. The shock was all the greater when the fresh discovery was made one day that another unmarried woman in the house, who was also a valued servant, was in the same condition. The two unhappy women had kept their secret from every one except from each other until it could be kept no longer, and they consulted together and determined to confess it to their mistress and abide the consequences.

Who were the men? was the first question asked. There was only one—Robert Coombe, the shepherd, who lived at the farm-house, a slow, silent, almost inarticulate man, with a round head and flaxen hair; a bachelor of whom people were accustomed to say that he would never marry because no woman would have such a stolid, dull-witted fellow for a husband. But he was a good shepherd and had been many years on the farm, and it was altogether a terrible business. Forthwith the farmer got out his horse and rode to the downs to have it out with the unconscionable wretch who had brought that shame and trouble on them. He found him sitting on the turf eating his midday bread and bacon, with a can of cold tea at his side, and getting off his horse he went up to him and damned him for a scoundrel and abused him until he had no words left, then told his shepherd that he must choose between the two women and marry at once, so as to make an honest woman of one of the two poor fools; either he must do that or quit the farm forthwith.

Coombe heard in silence and without a change in his countenance, masticating his food the while and washing it down with an occasional draught from his can, until he had finished his meal; then taking his crook he got up, and remarking that he would 'think of it' went after his flock.

195

The farmer rode back cursing him for a clod; and in the evening Coombe, after folding his flock, came in to give his decision, and said he had thought of it and would take Jane to wife. She was a good deal older than Ellen and not so good-looking, but she belonged to the village and her people were there, and everybody knowed who Jane was an' she was an old servant an' would be wanted on the farm. Ellen was a stranger among them, and being only a dairymaid was of less account than the other one.

So it was settled, and on the following morning Ellen, the rejected, was told to take up her traps and walk.

What was she to do in her condition, no longer to be concealed, alone and friendless in the world? She thought of Mrs. Poole, an elderly woman of Winterbourne Bishop, whose children were grown up and away from home, who when staying at Bower Chalk some months before had taken a great liking for Ellen, and when parting with her had kissed her and said: 'My dear, I lived among strangers too when I were a girl and had no one of my own, and know what 'tis.' That was all; but there was nobody else, and she resolved to go to Mrs. Poole, and so laden with her few belongings she set out to walk the long miles over the downs to Winterbourne Bishop where she had never been. It was far to walk in hot August weather when she went that sad journey, and she rested at intervals in the hot shade of a furze-bush, haunted all day by the miserable fear that the woman she sought, of whom she knew so little, would probably harden her heart and close her door against her. But the good woman took compassion on her and gave her shelter in her poor cottage, and kept her till her child was born, in spite of all the women's bitter tongues. And in the village where she had found refuge she remained to the end of her life, without a home of her own, but always in a room or two with her boy in some poor person's cottage. Her life was hard but not unpeaceful, and the old people, all dead and gone now, remembered Ellen as a very quiet, staid woman who worked hard for a living, sometimes at the wash-tub, but mostly in the fields, hay-making and harvesting and at other times weeding, or collecting flints, or with a spud or sickle extirpating thistles in the pasture-land. She worked alone or with other poor women, but with the men she had no friendships; the sharpest women's eyes in the village could see no fault in her in this respect; if it had not been so, if she had talked pleasantly with them and smiled when addressed by them, her life would have been made a burden to her. She would have been often asked who her brat's father was. The dreadful experience of that day, when she had been cast out and was alone in the world, when, burdened with her unborn child, she had walked over the downs in the hot August weather, in anguish of apprehension, had sunk into her soul. Her very nature was changed, and in a man's presence her blood seemed frozen, and if spoken to she answered in monosyllables with her eyes on the earth. This was noted, with the result that all the village women were her good friends; they never reminded her of her fall, and when she died still young they grieved for her and befriended the little orphan boy she had left on their hands.

He was then about 11 years old, and was a stout little fellow with a round head and flaxen hair like his father; but he was not so stolid and not like him in character at all events his old widow in speaking of him to me said that never in all his life did he do one unkind or unjust thing. He came from a long line of shepherds, and shepherding was perhaps almost instinctive in him; from his earliest boyhood the tremulous bleating of the sheep and half-muffled clink of the copper bells and the sharp bark of the sheep-dog had a strange attraction for him. He was always ready when a boy was wanted to take charge of a flock during a temporary absence of the shepherd, and eventually, when only about 15, he was engaged as undershepherd, and for the rest of his life shepherding was his trade.

His marriage to Martha Bawcombe came as a surprise to the village, for though no one had any fault to find with Tommy Ierat there was a slur on him, and Martha, who was the finest girl in the place, might, it was thought, have looked for some one better. But Martha had always liked Tommy; they were of the same age and had been playmates in their childhood; growing up together their childish affection had turned to love, and after they had waited some years and Tommy had a cottage and seven shillings a week, Isaac and his wife gave their consent and they were married. Still they felt hurt at being discussed in this way by the villagers, so that when Ierat was offered a place as shepherd at a distance from home, where his family history was not known, he was glad to take it and his wife to go with him, about a month after her child was born.

The new place was in Somerset, thirty-five to forty miles from their native village, and Ierat as shepherd at the manor-house farm on a large estate would have better wages than he had ever had before and a nice cottage to live in. Martha was delighted with her new home—the cottage, the entire village, the great park and mansion close by, all made it seem like paradise to her. Better than everything was the pleasant welcome she received from the villagers, who looked in to make her acquaintance and seemed very much taken with her appearance and nice, friendly manner. They were all eager to tell her about the squire and his lady, who were young, and of how great an interest they took in their people and how much they did for them and how they were loved by everybody on the estate.

It happens, oddly enough, that I became acquainted with this same man, the squire, over fifty years after the events I am relating, when he was past 80. This acquaintance came about by means of a letter he wrote me in reference to the habits of a bird or some such small matter, a way in which I have become acquainted with scores—perhaps I should say hundreds—of persons in many parts of the country. He was a very fine man, the head of an old and distinguished county family; and ideal squire, and one of the few large landowners I have had the happiness to meet who was not devoted to that utterly selfish and degraded form of sport which consists in the annual rearing and sebsequent slaughter of a host of pheasants.

Now when Martha was entertaining half a dozen of her new neighbours who had come in to see her, and exhibited her baby to them and then proceeded to suckle it, they looked at one another and laughed, and one said, 'Just you wait till the lady at the mansion sees 'ee—she'll soon want 'ee to nurse her little one.'

What did they mean? They told her that the great lady was a mother too, and had a little sickly baby and wanted a nurse for it, but couldn't find a woman to please her.

Martha fired up at that. Did they imagine, she asked, that any great lady in the world with all her gold could tempt her to leave her own darling to nurse another woman's? She would not do such a thing—she would rather leave the place than submit to it. But she didn't believe it—they had only said that to tease and frighten her!

They laughed again, looking admiringly at her as she stood before them with sparkling eyes, flushed cheeks, and fine full bust, and only answered, 'Just you wait, my dear, till she sees 'ee.'

And very soon the lady did see her. The people at the manor were strict in their religious observances and it had been impressed on Martha that she had better attend at morning service on her first Sun-

day, and a girl was found by one of her neighbours to look after the
baby in the meantime. And so when Sunday came she dressed herself
in her best clothes and went to church with the others. The service
over, the squire and wife came out first and were standing in the path
exchanging greetings with their friends; then as the others came out
with Martha in the midst of the crowd the lady turned and fixed her
eyes on her, and suddenly stepping out from the group she stopped
Martha and said, 'Who are you?—I don't remember your face.'

'No, ma'am,' said Martha, blushing and curtsying. 'I be the new
shepherd's wife at the manor-house farm—we've only been here a few
days.'

The other then said she had heard of her and that she was nursing
her child, and she then told Martha to go to the mansion that afternoon as she had something to say to her.

The poor young mother went in fear and trembling, trying to
stiffen herself against the expected blandishments.

Then followed the fateful interview. The lady was satisfied that she
had got hold of the right person at last—the one in the world who
would be able to save her precious little one 'from to die,' the poor
pining infant on whose frail little life so much depended! She would
feed it from her full, healthy breasts and give it something of her own
abounding, splendid life. Martha's own baby would do very well—
there was nothing the matter with it, and it would flourish on 'the
bottle' or anything else, no matter what. All she had to do was to go
back to her cottage and make the necessary arrangements, then come
to stay at the mansion.

Martha refused, and the other smiled; then Martha pleaded and cried
and said she would never never leave her own child, and as all that
had no effect she was angry, and it came into her mind that if the
lady would get angry too she would be ordered out and all would be
over. But the lady wouldn't get angry, for when Martha stormed she
grew more gentle and spoke tenderly and sweetly, but would still have
it her own way, until the poor young mother could stand it no
longer, and so rushed away in a great state of agitation to tell her
husband and ask him to help her against her enemy. But Tommy took
the lady's side, and his young wife hated him for it, and was in despair
and ready to snatch up her child and run away from them all, when
all at once a carriage appeared at the cottage, and the great lady herself, followed by a nurse with the sickly baby in her arms, came in.

199

She had come, she said very gently, almost pleadingly, to ask Martha to feed her child once, and Martha was flattered and pleased at the request, and took and fondled the infant in her arms, then gave it suck at her beautiful breast. And when she had fed the child, acting very tenderly towards it like a mother, her visitor suddenly burst into tears, and taking Martha in her arms she kissed her and pleaded with her again until she could resist no more; and it was settled that she was to live at the mansion and come once every day to the village to feed her own child from the breast.

Martha's connexion with the people at the mansion did not end when she had safely reared the sickly child. The lady had become attached to her and wanted to have her always, although Martha could not act again as wet nurse, for she had no more children herself. And by and by when her mistress lost her health after the birth of a third child and was ordered abroad, she took Martha with her, and she passed a whole year with her on the Continent, residing in France and Italy. They came home again, but as the lady continued to decline in health she travelled again, still taking Martha with her, and they visited India and other distant countries, including the Holy Land; but travel and wealth and all that the greatest physicians in the world could do for her, and the tender care of a husband who worshipped her, availed not, and she came home in the end to die; and Martha went back to her Tommy and the boy, to be separated no more while their lives lasted.

The great house was shut up and remained so for years. The squire was the last man in England to shirk his duties as landlord and to his people whom he loved, and who loved him as few great landowners are loved in England, but his grief was too great for even his great strength to bear up against, and it was long feared by his friends that he would never recover from his loss. But he was healed in time, and ten years later married again and returned to his home, to live there until nigh upon his ninetieth year. Long before this the Ierats had returned to their native village. When I last saw Martha, then in her eighty-second year, she gave me the following account of her Tommy's end.

He continued shepherding up to the age of 78. One Sunday, early in the afternoon, when she was ill with an attack of influenza, he came home, and putting aside his crook said, 'I've done work.'

'It's early,' she replied, 'but maybe you got the boy to mind the sheep for you.'

'I don't mean I've done work for the day,' he returned. 'I've done for good—I'll not go with the flock no more.'

'What be saying?' she cried in sudden alarm. 'Be you feeling bad—what be the matter?'

'No, I'm not bad,' he said. 'I'm perfectly well, but I've done work;' and more than that he would not say.

She watched him anxiously but could see nothing wrong with him; his appetite was good, he smoked his pipe, and was cheerful.

Three days later she noticed that he had some difficulty in pulling on a stocking when dressing in the morning, and went to his assistance. He laughed and said, 'Here's a funny thing! You be ill and I be well, and you've got to help me put on a stocking!' and he laughed again.

After dinner that day he said he wanted a drink and would have a glass of beer. There was no beer in the house, and she asked him if he would have a cup of tea.

'Oh, yes, that'll do very well,' he said, and she made it for him.

After drinking his cup of tea he got a footstool, and placing it at her feet sat down on it and rested his head on her knees; he remained a long time in this position so perfectly still that she at length bent over and felt and examined his face, only to discover that he was dead.

And that was the end of Tommy Ierat, the son of Ellen. He died, she said, like a baby that has been fed and falls asleep on its mother's breast.

XXV

LIVING IN THE PAST

DURING OUR frequent evening talks, often continued till a late hour,
it was borne in on Caleb Bawcombe that his anecdotes of wild creatures interested me more than anything else he had to tell; but in spite
of this, or because he could not always bear it in mind, the conversation almost invariably drifted back to the old subject of sheep, of
which he was never tired. Even in his sleep he does not forget them;
his dreams, he says, are always about sheep; he is with the flock, shift-
ing the hurdles, or following it out on the down. A troubled dream,
when he is ill or uneasy in his sleep is invariably about some difficulty
with the flock; it gets out of his control, and the dog cannot understand him or refuses to obey when everything depends on his instant
action. The subject was so much to him, so important above all others,
that he would not spare the listener even the minutest details of the
shepherd's life and work. His 'hints on the construction of sheep-folds'
would have filled a volume; and if any farmer had purchased the book
he would not have found the title a misleading one and that he had
been defrauded of his money. But with his singular fawn-like face and
clear eyes on his listener it was impossible to fall asleep, or even to let
the attention wander; and incidentally even in his driest discourse
there were little bright touches which one would not willingly have
missed.

About hurdles he explained that it was common for the downland
shepherds to repair the broken and worn-out ones with the long
woody stems of the bithy-wind from the hedges; and when I asked
what the plant was he described the wild clematis or traveller's-joy;
but those names he did not know—to him the plant had always been
known as *bithywind* or else *Devil's guts*. It struck me that bithywind

might have come by the transposition of two letters from withybind, as if one should say flutterby for butterfly, or flagondry for dragonfly. Withybind is one of the numerous vernacular names of the common convolvulus. Lilybind is another. But what would old Gerarde, who invented the pretty name of traveller's-joy for that ornament of the wayside hedges, have said to such a name as Devil's guts?

There was, said Caleb, an old farmer in the parish of Bishop who had a peculiar fondness for this plant, and if a shepherd pulled any of it out of one of his hedges after leafing-time he would be very much put out; he would shout at him, 'Just you leave my Devil's guts alone or I'll not keep you on the farm.' And the shepherds in revenge gave him the unpleasant nickname of 'Old Devil's Guts,' by which he was known in that part of the country.

As a rule, talk about sheep, or any subject connected with sheep, would suggest something about sheep-dogs—individual dogs he had known or possessed, and who always had their own character and peculiarities, like human beings. They were good and bad and indifferent; a really bad dog was a rarity; but a fairly good dog might have some trick or vice or weakness. There was Sally, for example, a stump-tail bitch, as good a dog with sheep as he ever possessed, but you had to consider her feelings. She would keenly resent any injustice from her master. If he spoke too sharply to her, or rebuked her unnecessarily for going a little out of her way just to smell at a rabbit burrow, she would nurse her anger until an opportunity came of inflicting a bite on some erring sheep. Punishing her would have made matters worse: the only way was to treat her as a reasonable being and never to speak to her as a dog—a mere slave.

Dyke was another dog he remembered well. He belonged to old Shepherd Matthew Titt, who was head-shepherd at a farm near Warminster, adjacent to the one where Caleb worked. Old Mat and his wife lived alone in their cottage out of the village, all their children having long grown up and gone away to a distance from home, and being so lonely 'by their two selves' they loved their dog just as others love their relations. But Dyke deserved it, for he was a very good dog. One year Mat was sent by his master with lambs to Weyhill, the little village near Andover, where a great sheep-fair is held in October every year. It was distant over thirty miles but Mat though old was a strong man still and greatly trusted by his master. From this journey he returned with a sad heart, for he had lost Dyke. He had disappeared one night while they were at Weyhill. Old Mrs. Titt cried for him as she would have cried for a lost son, and for many a long day they went about with heavy hearts.

Just a year had gone by when one night the old woman was roused from sleep by loud knocks on the window-pane of the living-room below. 'Mat! Mat!' she cried, shaking him vigorously, 'wake up—old Dyke has come back to us!' 'What be you talking about?' growled the old shepherd. 'Lie down and go to sleep—you've been dreaming.' ' 'Taint no dream; 'tis Dyke—I know his knock,' she cried, and getting up she opened the window and put her head well out, and there sure enough was Dyke, standing up against the wall and gazing up at her, and knocking with his paw against the window below.

Then Mat jumped up, and going together downstairs they unbarred the door and embraced the dog with joy, and the rest of the night was spent in feeding and caressing him, and asking him a hundred questions, which he could only answer by licking their hands and wagging his tail.

It was supposed that he had been stolen at the fair, probably by one of the wild, little, lawless men called 'general dealers,' who go flying about the country in a trap drawn by a fast-trotting pony; that he had been thrown, muffled up, into the cart and carried many a mile away, and sold to some shepherd, and that he had lost his sense of direction. But after serving a stranger a full year he had been taken with sheep to Weyhill Fair once more, and once there he knew where he was, and had remembered the road leading to his old home and master, and making his escape had travelled the thirty long miles back to Warminster.

The account of Dyke's return reminded me of an equally good story of the recovery of a lost dog which I heard from a shepherd on the Avon. He had been lost over a year, when one day the shepherd, being out on the down with his flock, stood watching two drovers travelling with a flock on the turnpike road below, nearly a mile away, and by and by hearing one of their dogs bark he knew at that distance that it was his dog. 'I haven't a doubt,' he said to himself, 'and if I know his bark he'll know my whistle.' With that he thrust two fingers in his mouth and blew his shrillest and longest whistle, then waited the result. Presently he spied a dog, still at a great distance, coming swiftly towards him; it was his own dog, mad with joy at finding his old master.

Did ever two friends, long sundered by unhappy chance, recognize each other's voices at such a distance and so come together once more!

Whether the drovers had seen him desert them or not, they did not follow to recover him, nor did the shepherd go to them to find out how they had got possession of him; it was enough that he had got his dog back.

No doubt in this case the dog had recognized his old home when taken by it, but he was in another man's hands now, and the habits and discipline of a life made it impossible for him to desert until that old, familiar, and imperative call reached his ears and he could not disobey.

Then (to go on with Caleb's reminiscences) there was Badger, owned by a farmer and worked for some years by Caleb—the very best stump-tail he ever had to help him. This dog differed from others in his vivacious temper and ceaseless activity. When the sheep were feeding quietly and there was little or nothing to do for hours at a time, he would not lie down and go to sleep like any other sheep-dog, but would spend his vacant time 'amusing of hisself' on some smooth slope where he could roll over and over; then run back and roll over again and again, playing by himself just like a child. Or he would chase a butterfly or scamper about over the down hunting for large white flints, which he would bring one by one and deposit them at his master's feet, pretending they were something of value and greatly enjoying the game. This dog, Caleb said, would make him laugh every day with his games and capers.

When Badger got old his sight and hearing failed; yet when he was very nearly blind and so deaf that he could not hear a word of command, even when it was shouted out quite close to him, he was still kept with the flock because he was so intelligent and willing. But he was too old at last; it was time for him to be put out of the way. The farmer, however, who owned him, would not consent to have him shot, and so the wistful old dog was ordered to keep at home at the farm-house. Still he refused to be superannuated, and not allowed to go to the flock he took to shepherding the fowls. In the morning he would drive them out to their run and keep them there in a flock, going round and round them by the hour, and furiously hunting back the poor hens that tried to steal off to lay their eggs in some secret place. This could not be allowed, and so poor old Badger, who would have been too miserable if tied up, had to be shot after all.

These were always his best stories—his recollections of sheep-dogs, for of all creatures, sheep alone excepted, he knew and loved them best. Yet for one whose life had been spent in that small isolated village and on the bare down about it, his range was pretty wide, and it even included one memory of a visitor from the other world. Let him tell it in his own words.

'Many say they don't believe there be such things as ghosties. They niver see'd 'n. An' I don't say I believe or disbelieve what I hear tell. I warn't there to see. I only know what I see'd myself: but I don't say that it were a ghostie or that it wasn't one. I was coming home late one night from the sheep; 'twere close on 'leven o'clock, a very quiet night, with moon sheen that made it a'most like day. Near th' end of the village I come to the stepping-stones, as we call'n, where there be a gate and the road, an' just by the road the four big white stones for people going from the village to the copse an' the down on t'other side to step over the water. In winter 'twas a stream there, but the water it dried in summer, and now 'twere summer-time and there wur no water. When I git there I see'd two women, both on 'em tall, with black gowns on, an' big bonnets they used to wear; an' they were standing face to face so close that the tops o' their bonnets wur a'most touching together. Who be these women out so late? says I to myself. Why, says I, they be Mrs. Durk from up in the village an' Mrs. Gaarge Durk, the keeper's wife down by the copse. Then I thought I know'd how 'twas: Mrs. Gaarge, she'd a been to see Mrs. Durk in the village, and Mrs. Durk she were coming out a leetel way with her, so far as the stepping-stones, and they wur just having a last leetel talk before saying Good night. But mind, I hear'd no talking when I passed'n. An' I'd hardly got past 'n before I says, Why, what a fool be I! Mrs. Durk she be dead a twelvemonth, an' I were in the churchyard and see'd her buried myself. Whatever be I thinking of? That made me stop and turn round to look at 'n agin. An' there they was just as I see'd 'n at first—Mrs. Durk who was dead a twelvemonth, an' Mrs. Gaarge Durk from the copse, standing there with their bonnets a'most touching together. An' I couldn't hear nothing—no talking, they were so still as two posties. Then something came over me like a tarrible coldness in the blood and down my back an' I were afraid, and turning I runned faster than I ever runned in my life, an' never stopped—not till I got to the cottage.'

It was not a bad ghost story: but then such stories seldom are when coming from those who have actually seen, or believe they have seen, an immaterial being. Their principal charm is in their infinite variety; you never find two real or true ghost stories quite alike, and in this they differ from the weary inventions of the fictionist.

But invariably the principal subject was sheep.

'I did always like sheep,' said Caleb. 'Some did say to me that they couldn't abide shepherding because of the Sunday work. But I always

said, Someone must do it; they must have food in winter and water in summer, and must be looked after, and it can't be worse for me to do it.'

It was on a Sunday afternoon, and the distant sound of the church bells had set him talking on this subject. He told me how once, after a long interval, he went to the Sunday morning service in his native village, and the vicar preached a sermon about true religion. Just the going to church, he said, did not make men religious. Out there on the downs there were shepherds who seldom saw the inside of a church, who were sober, righteous men and walked with God every day of their lives. Caleb said that this seemed to touch his heart because he knowed it was true.

When I asked him if he would not change the church for the chapel, now he was ill and his vicar paid him no attention, while the minister came often to see and talk to him, as I had witnessed, he shook his head and said that he would never change. He then added: 'We always

say that the chapel ministers are good men: some say they be better
than the parsons; but all I've knowd—all them that have talked to me
—have said bad things of the Church, and that's not true religion: I say
that the Bible teaches different.'

Caleb could not have had a very wide experience, and most of us
know Dissenting ministers who are wholly free from the fault he
pointed out; but in the purely rural districts, in the small villages where
the small men are found, it is certainly common to hear unpleasant
things said of the parish priest by his Nonconformist rival; and should
the parson have some well-known fault or make a slip, the other is apt
to chuckle over it with a very manifest and most unchristian delight.

The atmosphere on that Sunday afternoon was very still, and by
and by through the open window floated a strain of music; it was
from the brass band of the Salvationists who were marching through
the next village, about two miles away. We listened, then Caleb remarked: 'Somehow I never cared to go with them Army people. Many
say they've done a great good, and I don't disbelieve it, but there was
too much what I call—NOISE; if, sir, you can understand what I mean.'

I once heard the great Dr. Parker speak the word imagination, or, as
he pronounced it, im-mádge-i-ná-shun, with a volume of sound which
filled a large building and made the quality he named seem the biggest
thing in the universe. That in my experience was his loftiest oratorical
feat; but I think the old shepherd rose to a greater height when, after
a long pause during which he filled his lungs with air, he brought
forth the tremendous word, dragging it out gratingly, so as to illustrate
the sense in the prolonged harsh sound.

To show him that I understood what he meant very well, I explained
the philosophy of the matter as follows: He was a shepherd of the
downs, who had lived always in a quiet atmosphere, a noiseless world,
and from life-long custom had become a lover of quiet. The Salvation
Army was born in a very different world, in East London—the dusty,
busy, crowded world of streets, where men wake at dawn to sounds
that are like the opening of hell's gates, and spend their long strenuous days and their lives in that atmosphere peopled with innumerable
harsh voices, until they, too, acquire the noisy habit, and come at last
to think that if they have anything to say to their fellows, anything to
sell or advise or recommend, from the smallest thing—from a mackerel
or a cabbage or a penn'orth of milk, to a newspaper or a book or a picture or a religion—they must howl and yell it out at every passer-by.

208

And the human voice not being sufficiently powerful, they provide themselves with bells and gongs and cymbals and trumpets and drums to help them in attracting the attention of the public.

He listened gravely to this outburst, and said he didn't know exactly 'bout that, but agreed that it was very quiet on the downs, and that he loved their quiet. 'Fifty years,' he said, 'I've been on the downs and fields, day and night, seven days a week, and I've been told that it's a poor way to spend a life, working seven days for ten or twelve, or at most thirteen shillings. But I never seen it like that; I liked it, and I always did my best. You see, sir, I took a pride in it. I never left a place but I was asked to stay. When I left it was because of something I didn't like. I couldn't abide cruelty to a dog or any beast. And I couldn't abide bad language. If my master swore at the sheep or the dog I wouldn't bide with he—no, not for a pound a week. I liked my work, and I liked knowing things about sheep. Not things in books, for I never had no books, but what I found out with my own sense, if you can understand me.

'I remember, when I were young, a very old shepherd on the farm; he had been more 'n forty years there, and he was called Mark Dick. He told me that when he were a young man he was once putting the sheep in the fold, and there was one that was giddy—a young ewe. She was always a-turning round and round and round, and when she got to the gate she wouldn't go in but kept on a-turning and turning, until at last he got angry and, lifting his crook, gave her a crack on the head, and down she went, and he thought he'd killed her. But in a little while up she jumps and trotted straight into the fold, and from that time she were well. Next day he told his master, and his master said, with a laugh, "Well, now you know what to do when you gits a giddy sheep." Sometime after that Mark Dick he had another giddy one, and remembering what his master had said, he swung his stick and gave he a big crack on the skull, and down went the sheep, dead. He'd killed it this time, sure enough. When he tells of this one his master said, "You've cured one and you've killed one; now don't you try to cure no more," he says.

'Well, some time after that I had a giddy one in my flock. I'd been thinking of what Mark Dick had told me, so I caught the ewe to see if I could find out anything. I were always a tarrible one for examining sheep when they were ill. I found this one had a swelling at the back of her head; it were like a soft ball, bigger 'n a walnut. So I took my knife and opened it, and out ran a lot of water, quite clear; and when I let her go she ran quite straight, and got well. After that I did cure other giddy sheep with my knife, but I found out there were some I couldn't cure. They had no swelling, and was giddy because they'd got a maggot on the brain or some other trouble I couldn't find out.'

Caleb could not have finished even this quiet Sunday afternoon conversation, in the course of which we had risen to lofty matters, without a return to his old favourite subjects of sheep and his shepherding life on the downs. He was long miles away from his beloved home now, lying on his back, a disabled man who would never again follow a flock on the hills nor listen to the sounds he loved best to hear—the multitudinous tremulous bleatings of the sheep, the tinklings of numerous bells, and crisp ringing bark of his dog. But his heart was there still, and the images of past scenes were more vivid in him than they can ever be in the minds of those who live in towns and read books. 'I can see it now,' was a favourite expression of his when relating some incident in his past life. Whenever a sudden light, a kind of smile, came into his eyes, I knew that it was at some ancient memory, a touch of quaintness or humour in some farmer or shepherd he had known in the vanished time—his father, perhaps, or old John, or Mark Dick, or Liddy, or Dan'l Burdon, the solemn seeker after buried treasure.

After our long Sunday talk we were silent for a time, and then he uttered these impressive words: 'I don't say that I want to have my life again, because 'twould be sinful. We must take what is sent. But if 'twas offered to me and I was told to choose my work, I'd say, Give me my Wiltsheer Downs again and let me be a shepherd there all my life long.'

W. H. HUDSON'S
MAJOR PUBLICATIONS

The Purple Land that England Lost   1885

A Crystal Age   1887

Argentine Ornithology
  (With P. L. Sclater)   1888–1889

The Naturalist in La Plata   1892

Fan   1892

Idle Days in Patagonia   1893

Birds in a Village   1893

British Birds   1895

Birds in London   1898

Nature in Downland   1900

Birds and Man   1901

El Ombú   1902

Hampshire Days   1903

Green Mansions   1904

A Little Boy Lost   1905

The Land's End   1908

Afoot in England   1909

A Shepherd's Life   1910

Adventures among Birds   1913

Far Away and Long Ago   1918

Birds in Town and Village   1919

The Book of a Naturalist   1919

Birds of La Plata   1920

Dead Man's Plack and an Old Thorn   1920

A Traveller in Little Things   1921

A Hind in Richmond Park   1922

Rare Vanishing & Lost British Birds
  (With Linda Gardiner)   1923

INDEX